Business @ the Speed of Bots

THE AEIO YOU METHOD

HOW TO IMPLEMENT ROBOTIC PROCESS AUTOMATION THAT SCALES

IMPLEMENTING ROBOTIC PROCESS AUTOMATION IN A FAST AND SCALABLE WAY. GET READY FOR THE NEW AGE OF DIGITAL TRANSFORMATION

TONY WALKER

ISBN 9781702018395

© Copyright 2019 by Lean IA Ltd - All rights reserved.

The right of Antony (Tony) Walker to be identified as the author of the work in accordance with the Copyright, Designs and Patents Act 1988.

It is not legal to reproduce, duplicate, or transmit any part of this document in either electronic means or printed format. Recording of this publication is strictly prohibited.

FOREWORD BY
GUY KIRKWOOD
UIPATH'S CHIEF EVANGELIST

FOREWORD

As Tony says, "getting started with introducing new technology like RPA means winning the hearts and minds of key stakeholders and their teams." The most exciting moment in any RPA project is when you see the wide-eyed shock on a team member's face when they realise that they will never have to do that boring, mundane and repetitive task that has been part of their job for the past X years.

Tony's book is a bit like that; easily read, articulate and compelling. The AEIO YOU approach is nicely laid out and follows a logical sequence. Plus the interviews with leaders in the market add insights normally missing from this type of publication.

I've been saying for years that RPA is not rocket science (well, we actually work for NASA, so that bit is). What execs, team leaders and team members get from this book is clear instruction set on where to start, the right questions to ask, the community of the willing you need to build to capitalise on this amazing technology.

"Business is going to change more in the next five years than it has in the last twenty" (an adaptation of the first line of Bill Gate's book, Business @ the speed of thought)

Tony Walker

This book is dedicated to:

Mom "the maths teacher", Dad "the economist", little sis "the creative", baby sis "the adventurer"

Table of Contents

Is this book for you?	10
What is the book about?	11
IN THIS BOOK YOU WILL LEARN:	13
A: AWARE & ALIGN	15
Step 1 Understand the technology.	16
Are you an Early adopter?	18
What Centre of Excellence Team?	20
RPA: RP what? RP who? RP ay?	21
2 Know the Myths, Challenges and Benefits	25
3 Understand the Market	43
GUEST: SoftBot, New to the race	45
GUEST: Tabscanner	49
GUEST: Edge Tech	52
4 Choose the Right Solution	54
5 The Evangelist	57
6 Run a Pilot: POC or POV	61
Lots of ideas …in different directions	64
7 Centre(s) of Excellence	67
GUEST: Solutions Architect	80
Are you Aligned?	86
E: EDUCATE & EMPOWER	88
The most valuable resource of a business will always be its people	89
8 Bring in the Experts	90
GUEST: Broadgate Consultancy	91
9 Involvement breeds commitment	94
10 Valuable assets	97
Are they educated?	99
Scout culture	100
11 Upskill	103
Do they feel empowered?	105
I: INSPECT & IDEATE	105
12 EAR (Enterprise Automation Road Map)	114
Identify and Assess	117
13 Define and Measure	118
14 RPA suitability	119

15 Focus with the 80:20 rule	120
16 Process complexity	121
17 Prioritise	123
18 Zoom in further:	124
Inspect: Have you identified and prioritised opportunities?	131
Solution design is like herding cats	132
19 Root cause workshop	134
20 Solution design workshop	136
Ideate: Have your solutions involved all the right people?	137
Stakeholder management	138
O: OPTIMISE	140
21 Lean Thinking in Solution Design	140
As Is/To Be	152
Are your processes lean?	154
Roll out your automation solution	155
22 Business Cases: Cost, Benefits, and Sign Off	156
23 Process Definition Document (PDD)	161
24 Solution Design Document	166
25 Best practices and standards	168
26 Data preparation and test cases	173
27 Testing: user acceptance testing, debug, sign off	174
28 From live testing to support BAU	177
29 Launch	178
30 Reflect on your achievements	178
31 Scaling up: RPA factory (repeatable)	179
GUEST: RPA public speaker and author	181
Continuous Improvement	185
YOU	186
Y: YIELD	187
32 Realising the benefits	188
Are you getting value?	193
O: ORGANISE & OVERSEE	193
33 Maintain your benefits	199
34 Managing changes	205
Are you in control?	206
35 Circle back to the beginning: AWARENESS & EDUCATION.	207
U: UNCOVER, UPGRADE & UPSKILL	207
The 36th Step:	207
GUEST: Head of Automation and AI	208

Intelligent Automation, Analytics, and more	211
Artificial Intelligence	217
GUEST: Artificial Solutions	222
Cloud Computing	227
Super-charge your bots to do more	229
Have you identified ways to enhance your bots to gain more benefits?	231
GUEST: Operations Director	232
The 36 steps of the AEIO YOU method	235

Is this book for you?

Have you just started using Robotic Process Automation (RPA), are you looking to start up an automation Centre of Excellence (CoE) team in your company to leverage RPA and start building automation solutions, or perhaps you want your new CoE to mature and grow?

This book is for Head of Automation or Digital Transformation, RPA Managers and Change Management who have or are soon to bring automation into their organization and looking to set up a CoE.

Whether your current automation team is 2 or 20, understand the roles and responsibilities and set up of a good team. Identify which roles you may be missing, and what *scalable* framework your team can work to, in order to build an automation factory you can be proud of, which churns out solutions on demand. Also understand the behind the scene roles and considerations when it comes to maintaining your bots, things not mentioned as much in the media. Use this book as a guide to ensure you're using the **AEIO YOU method**

This book is for RPA Project Managers and Business Analysts who work in a CoE or Operational Excellence (OpEx) team and are responsible for delivering automation but are new to RPA.

Are your automation projects stalling or losing traction, or do you want to generate more opportunities and fill your pipeline?

Whether you've only just heard of robotic process automation yesterday, been doing it for a few months or a couple years, its very valuable to understand the entire lifecycle from identifying the problem, to designing, building and testing the solution, and supporting the new capability. With this knowledge you will be able to design and build much more robust 'robots' and intelligent automation solutions, and be able to boast much higher ROIs on your business cases. You'll be responsible for delivering much more benefits to your organizations or clients. Furthermore, you'll see how you can apply these same techniques and steps to implement advanced technologies like Artificial Intelligence. Use this book to check off each of the 36 steps of the **AEIO YOU method**

This book is also for COOs, CIOs and Operations Directors and RPA sponsors who want a comprehensive view of how RPA/automation is implemented

Read industry best practices and insights, to get high level steps on how to best implement Intelligent Automation. This with improve your awareness on what's been happening in the industry and what may be to come in the near future. This will help you understand the dos, don't, myths, challenges and benefits of automating your business processes, and give you a picture of what your team are doing …or should be doing. So, you can pass this book to them to ensure they are adopting the **AEIO YOU method**

What is the book about?

This book takes you through Lean IA's methodology which is a blend of best practices in RPA, change management and lean thinking.

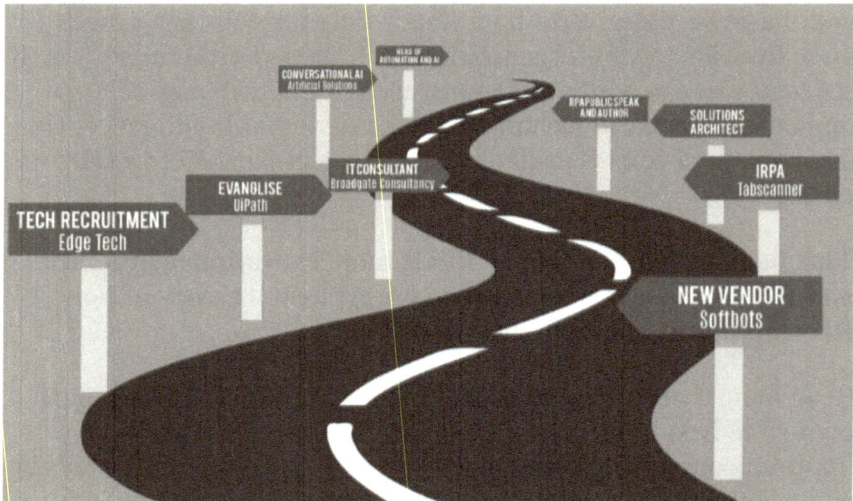

This 36 step AEIO YOU method was created by Tony Walker to guide automation teams around the world though the lifecycle of implementing new process automation technologies in the new fast paced world of digital transformation in a way that scales

Together we will walk along the digital transformation journey of implementing new automation technology, by going through the entire RPA lifecycle from idea to implementation to scalable intelligent automation, whilst we collect the insights and experiences of industry experts along the way. As we go further along this road we gradually get deeper into the RPA world.

You will notice that the AEIO YOU method used to bring RPA into your company can also be used for introducing any new technology, so we explore at the end how you can repeat these steps to bring Artificial intelligence into the fabric of your organization's business processes and teams

The underlying approach of AEIO YOU is to ensure you're applying Intelligent Automation to l*ean* (aka streamlined and optimised) business processes. Using lean thinking you can achieve a much higher ROI from your new technology than if you were to have just automated bad processes. Remember the old adage, GIGO (Garbage in, garbage out)

Though we will cover technical features and capabilities, **this is a non-technical, business focused view of RPA and intelligent automation** – looking at how your business can achieve things which have never before been possible with such ease, and how you can now explore avenues which previously would have been perceived as out of reach or unviable.

Intelligent Automation can enable a small company to make big moves like a large corporation, as they can service much higher numbers of customers, at greater speed, through various avenues in a much more affordable way, and without a drop in quality or customer experience.

Intelligent Automation allows large corporations to become more agile like a startup, as they will have a more flexible workforce which can be scaled up or down in minutes to meet demands. Their company will not feel as large as they will have insights and actionable information at their fingertips on what is happening throughout your entire organization without a slowdown from over cumbersome bureaucracy, and have a much more interconnected workforce, removing many silos that often stifle innovation and agility

IN THIS BOOK YOU WILL LEARN:

- The biggest mistakes and challenges most companies make and how to avoid them
- What is happening in the automation industry from perspectives of various industry business leaders, and a look at where it is heading
- How best to manage change through getting buy in from the top down, staying visible and keeping up momentum

- How companies late to the automation party or those struggling with scaling can accelerate their digital transformation past their competition
- A framework with tools, techniques and templates to build your 'RPA factory'
- How to get the most out of intelligent automation solutions, and how to repeat the 36 steps in the AEIO YOU method to discover and apply new intelligent technologies like AI
- At the end of the book is a glossary of TLAs and other industry acronyms and terms

IN EACH CHAPTER: We go step by step through AEIO YOU to ensure that you

1. Get the most out of this new and fast evolving technology which will make your teams and your business more efficient, allow you to move with agility like a startup and make big moves like a global corporation

2. I invite friends, and industry experts to share their experiences and insights, and share my own experiences

3. At the end of each chapter there are questions to ask yourself if you and your team have achieved each stage

A: Aware & Align

As new technologies and vendors enter the Intelligent Automation marketplace, it is important that you and your core team understand the differences in capability and the way each one can meet your business's unique needs

Once you understand the art of the possible, and that previously impossible avenues and new revenue streams are now open, you can renew your vision. Never has there been a better time to remove the silos in your organization and align departments and functions.

In the future, efficient organizations will be like living organisms, where data, information and resources flow freely and are shared without waste or unnecessary duplication – a single entity with symbiotic sub-systems, so it can grow, develop and respond quickly, fully aware of its surroundings.

Back in the post-automation age, data moved (or didn't) between business departments weighted by bureaucracy and politics. Start-up teams operated more fluidly and were much more agile. They could get to market quickly and adapt at light speed, leaving large slow corporations behind . Now, data can be exchanged at light speed and in the background, so corporations must align to achieve full digital transformation, otherwise they will be left behind or miss opportunities, like the HMVs vs Spotify, Blockbuster vs Netflix (2013) and Kodak vs Instagram.

Remember that technology levels the playing field. For years it's been cheaper and easier to start a global company than a local one, and it can reach millions of people for free. Workforces can be scaled up (or down) in no time, and accurate information is available in seconds without teams of analysts and experienced professionals needed in the past. With the right technology a small company can start tomorrow, take on a goliath business and win.

Intelligent Automation offers an exciting world of immense possibilities. We'll explore the various solutions available today and see how they can be applied in specific cases —cases that go beyond proof of concept.

The tech is amazing.

So why do companies struggle with it?

You've probably heard: **"50% of all RPA and AI projects fail"** and **"Only 1% of businesses have successfully scaled their RPA or AI"**…but if you've experienced the many angles of RPA and witnessed the many different industries that I have, you would know why.

So let's walk through the AEIO YOU step by step

Step 1 Understand the technology.

If artificial intelligence and machine learning are the brains of automation, then RPA (Robotic Process Automation) is the arms and legs

Thanks to Google's automation we no longer need to endlessly search through webpages and books. We can make contactless automated cash payment. Uber and Deliveroo have automated transport and delivery (sort of). Because these manual tasks have been removed from our lives, we can get things so much faster compared to 20 years ago.

In business RPA can remove technical roadblocks and mundane time-consuming tasks. It can fix the inefficient workflows required by rigid legacy systems— data migration, filling forms, 'copy & pasting' data between systems, and other menial tasks.

Automating tasks makes teams more productive. Staff can achieve targets and meet deadlines faster, freeing them up to focus on more interesting, creative, and value-adding work.

Fewer late nights can make teams happier, providing a better work-life balance, and heightened job satisfaction.

The business is happy because it can grow faster and create an enhanced customer experience.

RPA is software that is giving businesses massive power to better manage their workflows by developing faster, more efficient and accurate, teams. Early adopters to this technology are strengthening their competitive edge as it enables their business to run more cost effectively and greatly enhance services to customer. Now businesses can automate their processes in a matter of weeks or months, using existing systems, which is so much faster than cumbersome IT change that can take years

Are you an Early adopter?

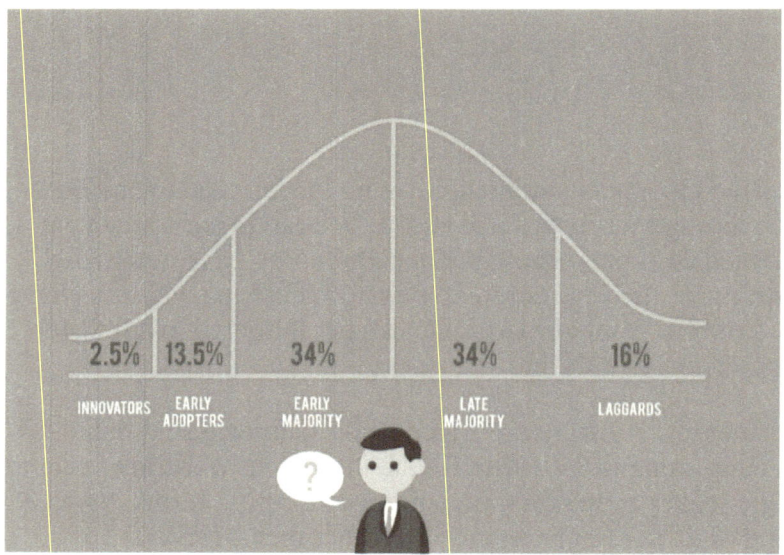

This adoption curve (a.k.a. the Rogers' bell curve) shows the general distribution of when people and businesses start to adopt a new technology. Change and the unknown is scary to humans, and so when something is brand new, be that a new technology or a new way of doing something, understandably very few people or businesses want to be the first to trial it.

The **Innovators** are the brave first few. They are the creators and probably those in the inner circle, or close enough to the new technology to appreciate it. They may be more educated in the new concept than the rest of the people. Generally, they are pioneering-risk takers with a passion for exploration

The **Early Adopters** are courageous and daring with a great eye for spotting the beginning of a winning trend. They are keen to stay on the cutting edge so they know when a new approach is starting to gain traction and they can capitalize on the potential upswing.

The **Early Majority** is pretty much everyone else who can really benefit from the new idea, the rest of the target market. They still can get an advantage over the remaining 50% of the population who have yet to discover it

The **Late Majority** are slow behemoth organizations that only move because their institutional investors are peering intently over their shoulder. By this time, they feel more at risk not to be involved, as they may get left behind, or may already be behind. Or perhaps they are smaller companies with not enough resources to research market changes and may have only caught wind of the new technology from observing a competitor.

The **Laggards** are frankly the old-fashioned, stuck-in-their-ways companies, who refuse to move, they believe their old way will prevail. Think Blockbuster.

So, if your company has started investing in robotic process automation technology, this is great news! Your company can see the massive potential of this competitive advantage. As of this publication (2019) you're probably at the latter end of the *Early Majority* phase. If you're also leveraging Artificial intelligence (such as Intelligent Character Recognition, Machine Learning or Intelligent Chatbots), then you're certainly in the *Early Adopter* camp, as your competitors will probably give it another 6–18 months to make this step.

By now I hope you've released a proof of concept (POC) and are setting up your Centre of Excellence (CoE) team to start compiling an automation catalogue of suitable processes ready to be prioritised.

Then you'll be able to add this into an Enterprise Automation Roadmap, giving your business a clear direction on what areas of your business you need to focus on and when.

What Centre of Excellence Team?

But maybe you're sitting there thinking 'What automation catalogue? What Centre of Excellence team?' If so, you're way behind many of your competitors. It's a sobering thought – especially if you're in manufacturing, banking, insurance, healthcare or utilities. You're even further behind if your company has lots of manually intensive processes, with a business model that is no longer scalable. Tell-tale signs are high attrition rates, highly frustrated employees, poor customer service, or large backlogs of back-office work which may be why you're hemorrhaging clients and money.

If you're in a company that isn't very far down the RPA road, and you're seeing these tell-tale signs, you most definitely have an excellent opportunity. Bringing RPA to your organisation will make a significant impact on your business's digital transformation when done properly.

RPA is still met with massive resistance and confusion, and it is surrounded by a lot of myths. Unfortunately, many companies have invested in RPA but their teams have only worked on proofs of concept, and now the rest of organization is trying to work out how to fit RPA into the overall strategy. This situation really is a messy gold rush with business leaders running around trying to make sense of it all. They all know RPA saves money, but few if any of them know how to make that happen. Only the savviest of companies and their rare-as-gold-dust RPA analysts are implementing RPA successfully and quickly increasing the gap between them and their competition.

If it's down to you to develop that overall strategy and own the enterprise automation roadmap, let's get you up to speed with the fundamentals:

RPA: RP what? RP who? RP ay?

Robotic process automation (RPA) is a software platform that can mimic a human's keystrokes and clicks on a computer screen. RPA works for almost any monotonous task your team does on their computer, for example, it could copy and paste information from one application to another. RPA allows companies to quickly develop 'robots' (bots) with minimal or no code.

Think of RPA bots as fast and accurate digital workers. They interact with the graphical user interface (GUI) of your company's legacy systems just as a human would. If the data inputs are standardized, these bots can follow any computer-based workflow that has pre-defined steps and rules-based decision points. If you can logically map out the process in a flowchart, then the bot can follow it. Think of it as an Excel macro on steroids. In fact, some of the underlying technology, such as screen-scraping and automated workflows, has been around for decades.

There is something that the bot can't do – understand unstructured data inputs like a free-text emails or comments. For those, you'll need to plug in some Machine Learning capabilities. Nor can the bot make subjective decisions because they require human intuition (not even AI has reached this level in a general sense). Your bots can either work alongside your team as virtual assistants to enhance productivity, accuracy and speed, or they can work in the background as an additional workforce which you can increase or decrease to match demand.

You can use these 8 characteristics of processes to find suitable tasks for automation:

1. The process **doesn't require human intuition**. Bots are ideal for those mind-numbing processes which don't require much thought.

    E.g., Copy/pasting, filling in forms, checking data, and so on.

2. You want a **stable process** that doesn't change much for different situations and has minimal exceptions. The process is one that is carried out exactly the same for almost every case, or there are only a handful of ways to handle a case.

E.g., Processing a payment or on-boarding a client

3. The process should be **rules based**, where any decision is logical: "If THIS then do THAT, ELSE do SOMETHING ELSE".

    E.g., If MARRIED send form A, else if SINGLE send form B

4. To get the most value for your efforts the process should have a **high volume** of cases passing through it.

    E.g., Transactions that require huge teams of people

5. The process needs to be **digital end-to-end** as the bots only exist on the computer.

    E.g., If the process also requires scanning paper or picking up the phone, the bot can do part of it, but a human has to intervene in order to complete it

6. The easiest processes to automate are those which are **manual and repetitive**. Human brains aren't as good with mundane processes as we tend to make mistakes.

7. **No pending changes** in the next 3 – 6 month on either the process or the underlying systems. Otherwise the automated process will need to be designed before any benefits are realised

8. The process must have **standardised input data**.

    E.g,. Information from a drop-down menu or a selector

Here's the difference between standard and non-standard data, which will demonstrate why the bot will not understand free text without any intelligence

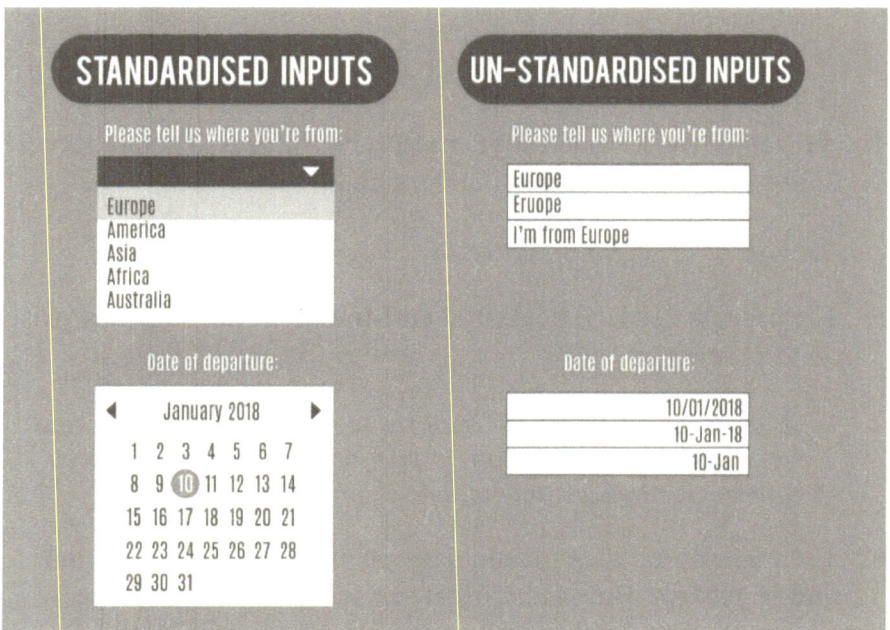

Because the bot will not understand free text without any intelligence it's important to know the difference. On the left are standard data inputs: where a person is from and their date of departure. On the right are non-standard inputs, or free text, for the same information. If the input fields allowed free text, the ways people can enter the same information could vary greatly. Unstructured data would be impossible for an RPA bot to process as it would need to be programmed to look for a plethora of different possible answers and would need to understand spelling mistakes too.

The early adopters of RPA were companies in manufacturing, banking, insurance, law, utilities and healthcare. A key reason RPA was created was that companies' realization that a fast and complete overhaul of their old systems was just too expensive and cumbersome. The realization started as a domino effect. Customers demanded faster, cheaper services, so IT departments became overwhelmed trying to address all the shortfalls in the required timeframe. Then companies rapidly became more technology hungry to keep up with customers' demands. To make matters worse, technology grads had learned new systems and new programming languages, so the resource pool of people who could modify legacy systems was drying up. Another important reason for RPA was the decline in labour arbitrage, a method in which companies outsource work to offshore staff as a way to keep up with demand because staff costs were cheaper overseas.

So, RPA put the power of automation into the hands of the business units, so they could optimise their teams, make big savings, and increase their speed and scalability. As a result, companies had more opportunities to provide more convenient and accurate services, improving the customer experience.

Nowadays RPA is so much more as it continues to advance and encompass more intelligent capabilities. With low- or even no-code, users can drag-and-drop functionality into a workflow and immediately make big changes in their teams

…But wait, let's take a look at the pitfalls and consequences of a lack of control, if RPA were democratized freely.

2 Know the Myths, Challenges and Benefits

"A myth is a way of making sense in a senseless world" – Rollo May

Whilst networking with business leaders and analysts in the RPA industry in London and engaging with those from India, Europe and the US, I heard the same questions and concerns again and again. In my investigation over the last few years on and offline there have been several themes for why businesses have been struggling, and the same misconceptions and messages have been circulating.

Let's debunk some of the key myths still in circulation which are the cause of much hesitation and resistance in many organisations.

Myths

- We need an army of bots.

    People thought that although automation technology was a lower-cost alternative to hiring a large workforce or upgrading a company's software, it was only available to the big corporations with deep pockets. Yes, there are infrastructure costs, and experts must be hired to get things going. However, when you buying licenses from RPA vendors you don't need to buy hundreds of bots for it to be worthwhile. In fact, it's not advisable, even for large organizations. At the start, you want to grow your digital workforce gradually.

    You can start with just one or two bots, and your business can make savings and scale up the workforce immediately. You could even start with zero bots. Many business models exist where you don't even need to own a bot, but you can rent a bot or acquire RPA as a Service (RaaS).

    Furthermore, some vendors offer free versions of their software that you can download onto your computer and start building as many bots as you want straight away. It's just like building a macro on Excel to automate a simple task. These bots can only run locally on the computer you've downloaded it on, but you can benefit immediately – as if you had a virtual assistant waiting for your commands.

- Bots will take all our jobs.

Job loss is by far the biggest fear in companies to date. Understandably no one wants to feel like their job is at risk; I had even read in the news how analysts were the most at risk as their roles were highly logical and data driven. As an analyst myself, this was a *delight* to hear. When your RPA team start going into different departments to find automation opportunities fear of job loss is the biggest challenge they will face. It's a very sensitive topic which needs to be handled carefully, otherwise it can stifle any momentum you were hoping to build.

Change management is vital to ensure the right messages are communicated but businesses need to be thinking about how their staff (their most value assets) can be upskilled and look at redesigning career roles and paths to align with digital transformation.

Bots should be seen as enablers to a workforce not substitutes because human workforce's experience and intellect is too valuable to lose. In many circumstances RPA has seen an increase in jobs as teams become more productive and cause companies to grow. New technologies always give rise to new types of jobs. Bots can remove the burden of tasks from the workforce, but this gives businesses the opportunity to upskill their staff – by attending specialist courses. I once heard the phrase that 'RPA should take the robot out of the human' so that the human can do more interesting and creative work.

- Ai bots can do most human tasks.

There's a lot of scaremongering about a Terminator-style uprising where AI becomes as smart as, if not smarter than humans and takes over. There was even a recent debate (August 2019) where Jack Ma and Elon Musk debated the benefits or threats of AI becoming smarter than humans.

Whatever the theories about what AI could achieve, what we know now is that AI is miles away from being smart enough to replace humans. Yes in March 2018, an AI system was able to read handwriting faster and more accurately that humans could, and for other specific tasks (like playing chess or Go), AI can out performance humans. AI can even understand natural language, it can learn, and it can recognize images, but it still is a fair way from replicating human intuition and reasoning.

Many leaders in technology believe we are effectively Cyborgs, fully dependent on our technology. Your mobile phone is pretty much a prosthesis as many of us could not exist without it for more than a few waking hours. However, technology as always been an enabler to allow us to move through life at lightning speed compared to 30 years ago.

Intelligent automation and AI bots are no different. They are virtual assistants that improve staff productivity and enhance their careers. These intelligent bots are owned by their team (not by IT) to do the lower straightforward work, leaving the higher-level thinking to us humans. Bots are there to enhance our efficiency and effectiveness, not replace it.

- RPA means the end of BPO

  Business process outsourcing (BPO), generally to overseas companies, was a way for companies to reduce their staff costs by about 50%; however, an RPA bot is about 10% the cost of an on-shore full-time employee. Then salaries overseas increased as the economy improved in the low-income countries and low-income areas within those countries. As a result, RPA was introduced as a countermeasure to maintain savings. However, BPO companies have also advanced and upskilled to provide digital BPO, where they can build and manage your bots – we'll examine digital BPO shortly.

- Bots don't make mistakes

Error-free bots are one of the biggest misconceptions and a core reason many business leaders have seen sub-par results in their RPA initiatives. At Lean Intelligent Automation we say, 'Garbage in, garbage out'. If your bot has been designed poorly, it doesn't manage data integrity well, and has poor error handling, then it will make mistakes. In other words, a bot does exactly what it's told; it will not do what it was intended to do if the right precautions have not been considered. It's not like a human who will make a mistake and then realise it. If a bot is programmed incorrectly it will carry out a task incorrectly 1000 times.

Preferably a bot should be programmed to do only a specific thing. Any deviation from that thing should be passed to a human to do because it's difficult or impossible to programme the bot to handle every exception. Certain unforeseen exceptions, like pop-up windows, or bad data inputs could cause the bot to complete the job incorrectly or not at all. So if the data doesn't match the criteria, or something doesn't match the predefined process steps, then the bot should reject it to avoid making any mistakes

- RPA will almost make IT department redundant.

    RPA was created to elevate the demands for assistance from the IT department, as IT staff are already inundated with other technology demands from the business. Automation of business processes will by no means replace what IT teams do. It's vital to include the IT department early on to ensure you have the right infrastructure set up to manage your robots. RPA is **heavily** dependent on the IT team, as your developers will need to have development and production environments for building, testing and deploying your robots.

- Automation is the destination.

Simply, no. Automating processes just because everyone else is doing it definitely wrongheaded and can lead to knee-jerk reactions. I've met with many companies who chose the most complicated tasks to start automating just because someone shouted about those tasks the loudest. Starting with a complex task could stall your whole operation. Stakeholders will soon lose hope and interest due to the RPA team taking too long or being unsuccessful at automating a process which was unsuitable. The aim is for better ROI, accuracy and flexibly.

Challenges

- Missing or unavailable data

    When it comes to running successful RPA programmes, data is key. You want to measure whether the opportunities you're exploring are worthwhile, and you also want to prove that benefits from the change have been realised.

    Missing or inaccurate data gives a false view on the potential savings or return on investment of your automation incentives. You want to measure the weekly volumes and average handling times for each process as accurately as possible, so you can calculate the amount of FTE (full-time employees) effort these processes use, which will allow you to start prioritizing your opportunities.

    It's best to collect sample data and observe the process to sense-check the metrics. Some teams don't collect data like weekly volumes or AHT (average handling times), and sometimes it's difficult to extract data from the system. Also, unless you're in manufacturing, it's generally frowned upon to walk around with a stopwatch to time staff on how long their processes take to complete. To get a rough idea of AHT your team may have no other alternative but to ask several subject matter experts (SMEs) for their gut feel, however, recording the screen as staff carry out the process would be better

- Staff's resistance to change

    One of the biggest challenges is resistance from staff who may be directly affected by the change. If they can't see how it benefits them, why should they be interested? If they feel their job could be at risk, why would they want to cooperate? A good communications plan can mitigate the resistance, and conveying opportunities to upskill or re-deploy staff alleviates fears of being replaced and shows staff they are valued.

    Getting buy-in should be first on the agenda and maintained continuously. You can keep staff involved with regular lunch-and-learns or periodic workshops that makes staff aware of the technology and shows them how to get involved. Keeping the door of your Centre of Excellence open and continuing to communicate with staff can relieve many tensions about introducing automation technology.

- Loss of traction

    It's great if you have made a start by rolling out some POCs but perhaps things have gone cold and stakeholders have lost interest. If you've lost traction it could be because you've made the same mistake that countless other businesses make: you've tried to solve a really big problem or your team have chosen a very complicated process. These opportunities take far too long to deliver. When starting up, you need to deliver some quick wins regularly, mixed with some good-sized opportunities to get the momentum going. I'll show you how to choose the right mix of processes to start with and how to prioritise the rest

- No clear governance

If there are no checkpoints or no gated process, and roles and responsibilities have not clearly been defined, then no one is accountable for each step of the progress. Your project is at risk of going around in circles or halting all together. Having clear, written governance that is agreed by all parties upfront, and with senior sign off, is the only way to avoid this pitfall.

Not only do you need to know who is responsible for each part of the implementation process, there needs to be an agreed turnaround time for each step. Assemble questions such as 'How long should it take to get robot access to applications, or for the Process Definition Document (PDD) or Solution Design Document (SDD) to be signed off?' and 'How long will it take for the bot to be handed over from the developer to the support team?' Also consider that at each checkpoint or gate you need clear acceptance criteria.

- Process clarity

    The PDD is the most important document in this whole process. This is what translates a business problem, into a technical solution. RPA business analysts must manage the relationship between the SME & developer, so that the developer creates a bot that does what the business wants.

    Its important that the process steps are well documented at a detailed level, showing the clicks and keystrokes the robot should make. The developer will most likely be unfamiliar with the process so there must be no gaps or jumps between one step and the next.

    This document must include everything the developer needs to effectively become an immediate expert of the process. The developer needs to know:

    o Where the bot takes the information from

- How it will receive it and in what format (e.g., what is the folder location, website address or email address, and will it have a specific subject title? Will the file be in xlsx or pdf format?)

- The precise format of the data (e.g,. what type of data is in each field, what is the file naming conventions are, and other data integrity information).

- What the bot must do with the information

- Where or to whom to send it.

- What time(s) the process needs to run, what days, and if there are any service level agreements (SLAs). This may determine whether multiple bots might be needed to process all cases in time to meet the deadline.

We will take a closer look at the PDD document structure later in the book.

- Documentation can differ from reality.

This links to the previous point. A major faux pas of your RPA team would be to rely on work instructions and other process documentation to design the automated process. What's written down tends to be quite different to what actually happens. There are 2 simple reasons for this:

1. Work instructions are usually written by the experts who know the process inside out. This is susceptible to 'the curse of knowledge', which means assumptive leaps may exist between steps, and the author may automatically believe this to be obvious.

2. The team or individuals might have found potentially better way(s) to do the process after the instructions were written. In lean thinking, the best way should be the standard, so maybe it's time for the team to find out which team member is the most efficient and use that person's knowledge to update the formal process documents.

To avoid this pitfall, it's always recommended to walk through the process with the SME. Shadow the person to observe how the process is executed in practice, so that your RPA analyst can fill in any gaps identified in the documentation.

- Test vs. Live (Production) Environments

More often than people want to admit, bots can act differently in the live environments than they did during testing. There are several reasons for this but mainly it's due to the application versions in the test environment being slightly older to that in the live environment.

Even if you have the most updated version in test, the bot may still act unexpectedly for unknown reasons, hence why a two-week, live test period is advised (or longer if the process is a high-risk, high-impact process such as one involving finances). During the test period the developer and support team can watch the bot closely and immediately fix any bugs.

We'll look again at testing the process in both the test and live (or production) environments later on, but ensuring the test platforms have the most up-to-date versions to match the live environment as close as possible can minimize risks.

Benefits

Getting started with introducing new technology like RPA means winning the hearts and minds of key stakeholders and their teams, so communicating the benefits clearly and early is vital. It is also key to ensure that teams are sufficiently educated on RPA so that they understand what to expect and what's in it for them. A clear and well-thought-out communication plan is needed, so at a minimum it reduces resistance and aids a smooth rollout, and at best, it gets people excited.

RPA alleviates the mundane, time-consuming and repetitive tasks from the operations teams' day-to-day work. These high-volume, monotonous workflows are prone to human error, and they can distract your staff from focusing on the more interesting, value-adding parts of their roles.

Freeing up staff time gives team managers the opportunity to help staff develop their careers. Staff can also use the time to lean into aspects of their work that may involve more interaction with other colleagues and customers, and to work more with problem-solving and decision-making tasks.

If you're an employee who has noticed RPA coming into your company, this surely opens a window of opportunity to progress so reading this book will definitely get you ahead of the pack.

1) Cost savings

Companies have been outsourcing overseas for decades to reduce operating costs, but RPA can save even more, especially as labour arbitrage is in decline as global wages are on the rise. RPA cost-savings from reducing headcount has been the go-to benefit most businesses focus on, as it is the fastest and most direct way to affect the bottom line. RPA is faster to implement than an IT change. Bots cost less that a full-time employee, don't require training, don't ask for pensions and can work 24 hours non-stop. Overseas personnel are about half the cost of an onshore full-time employee, however an RPA digital worker reduces costs dramatically, so it's very apparent to why business leaders are so keen to use digital workers to expand their workforce.

That said, companies need to realise that the real value of their business is their staff, who have immense knowledge of their business, their domain and their customers. The worst mistake a business can make is to remove staff, only to rehire new staff some months later after the business has grown (with the added cost of re-training new hires who have minimal domain or business knowledge). What's very disappointing, as I've witnessed, is that many businesses don't even have a knowledge-retention plan, even for staff who leave organically, but that's another story.

2) Develop a highly productive and efficient team.

It's not just about getting about the same amount done with less. If your business is expanding, you can get even more done with the same staff. By passing these non-value-added tasks to a bot and removing technical roadblocks that exist in your legacy systems, your team can focus on adding value to your customers, improving customer service, and spending more time discovering ways to solve business problems and make better strategic decisions.

In addition, when your team has a bot which is tracking the work flowing through your process, management can get much better MI (management information) on how the team are doing as they interact with their digital team mate(s). Better insights can lead to continuous improvement of the team's performance

3) Enhanced customer experience with increased accuracy and speed

   Repetitive tasks are prone to human error as our brains aren't designs for that, whereas a bot will do exactly what you tell it, consistently and at scale.

   Taking 'the robot out of the human' eliminates human error from a process, avoiding potential financial, compliance or reputational risks. Removing error also greatly enhances the customer experience and satisfaction in two ways. The customer receives accurate information as requested at speed, and the human side of the workforce can dedicate more time and energy to attentive customer service

   This benefit is regularly overlooked. Right-first-time means customers don't need to call back to fix errors, answer-shop to get a second option or complain about incorrect information, and customer needs are addressed faster due to reduced queuing. RPA also allows more revenue-generated services to be available to customers through avenues which were previously seen as not financially viable.

4) Have a scalable and flexible workforce at your fingertips.

   If teams in your company experience seasonal demand, or if your company is growing and team managers are concerned about the future ramp up of work, you may have a problem. In the past, you might have hired a temporary workforce for a few busy months, or struggled to get funding to quickly hire a few extra team members.

You may not need either option with RPA. It allows you to immediately scale up or down your digital work force as you see fit, and it can to make life less stressful for your staff with fewer late nights and less overtime. The only additional cost might be to acquire more server space or licenses to accommodate the increased demand if your current bots are at full capacity. Furthermore, RPA can track volumes, handling times and other metrics to allow you to discover insights and forecast more accurately where and when these demand spikes will happen, so teams can be better prepared.

5) Compliance

Another form of saving is by ensuring teams remain compliant. Think GDPR. Many companies missed the deadline on this. By automating business-critical processes, your company can meet regulatory requirements on time, and reduce the risk of being fined.

From a business-compliance perspective, business rules can be embedded into the process during the process redesign stage, so that management can have more control of operations.

When businesses have look at the bottom-line, much of the focus has been on cost savings; however, cost avoidance (from fines, complaints, reputational damage) can put a real number against the benefits of automating these types of processes.

2B Mistakes

1) Lack of senior or business leader buy-in

Getting pro-active senior buy-in is the only way to get a successful RPA programme moving. During stakeholder engagements it's vital to get top-down enthusiasm and demonstrate how RPA can enable their senior management teams to be more efficient and cost effective.

If senior management are fully on-board, they can remove blockers, deploy the right personnel to assist the RPA team to identify the opportunities, and they can re-direct their limited funds towards more RPA initiatives. If senior leaders believe in the technology, they will be more likely to redirect funds away from tried-and-tested methods, such as recruitment or outsourcing, and start investing in this new and seemingly untested (in their area at least) technology.

2) Lack of time/commitment from SME

Though buy-in from senior leadership gets things moving, the second hurdle is getting the managers and the team themselves excited and keen to get involved. This is potentially the hardest sell, as many may be totally content with the old way of doing things. Some may not feel comfortable with change in general, and the rest may quietly fear that this new technology could change their job or replace them.

Once team managers and SMEs are on board, the right expectations need to be set with them so they understand how much time the RPA team requires from them in order to design the right solution, to the necessary detail. If expectations are not clear upfront, staff and managers may grow frustrated with intermittent disruptions to business as usual. It's important that they understand how the discovery and implementation process works, and the RPA team needs to clearly layout their method, and explain that it's not a one-and-done event, but a staged process.

Other mistakes to look out for:

3) Poor stakeholder education and communications

Doing RPA in a dark corner of the office leads to staff thinking the worst. Openness and visibility is best; take them on the journey with you. At a minimum, an easy way to stay visible is to create a newsletter and an online portal with information on the technology, the process and a discussion forum for questions and concerns.

In addition to the visibility provided by open-door policy and regular lunch-and-learns or periodic workshops, the Centre of Excellence should stay in communication with staff. Popping up and disrupting teams for a few days, then disappearing without results can be very irritating. Stakeholders in a few organizations I've worked with have expressed their frustrations about previous RPA teams who attempted to implement but were shut down or moved on to a different team without any explanation. Understandably when I came to launch the RPA properly I was initially meet by reservations and reluctance. Even if your RPA team needs to close down an initiative, its courtesy to inform the team as to why, so that the relationship ends on good terms and they are welcoming the next time you circle back.

4) Lack of IT ownership and understanding

IT needs to understand what RPA is and what role they play in making it work. The CoE team should also have close ties and open communication with the IT who will be providing the infrastructure, be that local or remote servers or something else.

It's also important to note that if a server goes down, it's out of the control of the support team and is in the hands of IT to get it up and running fast. An operational-level agreement between the two departments explaining how fast something must be fixed should be agreed upfront. Otherwise your SLA with your stakeholders could be impeded by something out of your control.

You'll even need to work with IT to create your business continuity plan in case their servers or the power goes down, and involve the operations team in reverting back to manual processing as a last resort.

5) Unclear governance process and responsibilities

You need to have controls or criteria at each checkpoint throughout the implementation lifecycle, and clear roles and responsibilities need to be assigned. Otherwise the lack of clarity will breed anarchy when you try to start to scale, and that's only if your team can succeed in pushing a bot through the door in such a state.

Having a clear framework and set of criteria for how an initiative will pass through each succeeding stage will ensure that time isn't wasted trying to implement difficult processes, accept badly coded robots or go around in circles to find who is or should be accountable for each step. Setting timeframes and escalation paths is also important, as not having defined turnaround times (for things like getting robot or developer access to applications, or email addresses for specific robots) can delay projects significantly. If something has been delayed, your team needs to know who these issue should be escalated to.

Keeping the pace is important so something as simple as a table to show who is Responsible, Accountable, who to Consult and who to Inform (a RACI table) can ensure everyone along the process knows what is expected of them.

6) Unrealistic expectations:

RPA has been hyped in the media for years, yet still only 50% of RPA and AI initiatives succeed, and only 1–2% of companies have managed to scale successfully. It's understandable that many business leaders are greatly disappointed by results.

RPA isn't a silver bullet. On its own it can only process structured data, and structured data only amounts to about 20% of the data available in your organization. Central to the survival of your automation programme is setting realistic expectations so that key senior stakeholders don't feel underwhelmed with results and lose faith. RPA even on its own has amazing potential as that 20% may save your company greatly in time, money, protect your reputation and improve customer experience. Trust in the technology, but don't believe the hype.

7) Choosing the wrong process

Time and time again I've heard of companies or consulted with companies who were in the midst of trying to automate a complex, low-volume, highly volatile process as one of their first processes. Sometimes the process even requires several forms of Ai. This is a clear sign of "shiny object syndrome" (SOS), the propensity to chase something because it's new and shiny and everyone else is doing it. Or perhaps a case of "The one who shouts loudest" gets their process automated. At this point the RPA sponsor or senior needs to step in. New things can have great value, but there must be a logical approach to utilitising such a powerful tool.

Choosing the wrong process to start with is guaranteed to stall an automation programme. Stakeholders will lose interest, the RPA team will get disheartened, and the RPA sponsor will be wondering whether she will ever get a return on the investment and may consider pulling the plug on the whole thing. Such a toxic perception of this tool will swiftly destroy what you are trying to achieve. If you suspect you have started with the wrong process, this realization may mark the last chance you have to hire an experienced RPA analyst or success manager to turn things around and get the process back on track.

If you have just hired an expert and have yet to choose your process, it would be wise to listen to the expert to avoid choosing badly.

8) Not streamlining process first

There are two schools of thought in RPA. Roll out automation fast leaving the processes as it is because it will immediately impact the bottom line. Or optimise a process first with lean thinking so that the process is redesigned to be streamlined and is designed for a robot instead of a human. At Lean IA, you can surely guess which one we prefer. Garbage in, garbage out.

Optimising a process can provide a much faster payback period as costs may be lower because the process may be designed to be simpler. Imagine a process where the human goes back and forth to pull data from one application into another. This workflow is long winded as the human has limited memory, whereas the process could be redesigned for the robot to take all the information from the whole page, then paste it into the other page in one go, thus eliminating the old back-and-forth way.

3 Understand the Market

RPA was created in 2003 when Blue Prism launched the first product 'Automate'[i], however the underlying functionalities (macros, screen-scrapping and recording) have been around since the 90s.

The three biggest RPA players currently are Automation Anywhere, Blue Prism and UiPath. The underlying code is the same, and they share similarities in parts; however, RPA has advanced considerably in the last few years. With the plethora of AI solutions which are now compatible with RPA vendors, there are a whole ecosystem available to you. It is a good idea to have a consultant who understands the different vendors and compatible plug-ins to match the right solution to your specific business needs.

As previously mentioned, RPA can only process 20% of the data as the rest is unstructured (non-standardised, free text). For this reason, in the recent years businesses have grown frustrated with the limitations of RPA and have been disappointed by the results. Consequently, many businesses new to automation have jumped straight to iRPA (intelligent RPA), building AI capabilities from conception to provide a unique offering. Leading RPA vendors have started to invest heavily in meeting this demand by including such capabilities as intelligent character recognition, Machine Learning and data analytics in their software ecosystems to address additional business needs. Some leading vendors have decided not to re-invent to wheel but have chosen to stick to their forte and instead partner with the top AI providers.

Currently more than 50 vendors are on the main stage, many with their own micro-ecosystem, however the AI market on the other hand is quickly becoming much more crowded as cloud computing and open source libraries makes the barriers to entry much lower. You may have noticed on LinkedIn and technology publications tonnes of new AI start-ups appearing and pitching new and innovative solutions to old problems. This growth is no surprise as a lot of money is flowing into these industries.

RPA: Global spend $2.4bn in 2022[1]

Gartner forecasts the RPA software market will grow by 41% year over year through 2022. Technology product managers must capitalize on this expected growth by developing tools that allow adopters to more easily scale their RPA deployments

Ai: Global spend $8.3tn by 2035[2]

---

[1] https://www.gartner.com/en/newsroom/press-releases/2018-11-13-gartner-says-worldwide-spending-on-robotic-process-automation-software-to-reach-680-million-in-2018

[2] https://emerj.com/ai-sector-overviews/valuing-the-artificial-intelligence-market-graphs-and-predictions/

Forester says the global Artificial intelligence business value will reach $1.2 trillion by 2020, and Accenture projects that by 2035 the value will reach $8.3 trillion in just the United States compared to $814 billion in United Kingdom, $2.1 trillion in Japan, and $1.1 trillion in Germany.

Within 18 months to 2 years, pure RPA solutions may disappear as these two worlds quickly converge and companies feel more comfortable adopting Ai.

GUEST: SoftBot, New to the race

*I reached out to our network to understand what the challenges and barriers of entry are to the intelligent automation market. I caught up with **Co-founder and CEO of SoftBot, Mark Beason,** to understand the gap they saw in the market and what they found so attractive in this space.*

***SoftBot**, is a new entrant to the RPA market that specialises in solving and packaging customer and employee automation solutions for the legal, insurance, call centre and mid-market manufacturing sectors problems.*

What attracted you personally to the RPA/intelligent automation industry?

I have been working in the Software industry for 30 years in automated delivery solutions, an area which is now defined as RPA. The rise in automation software has enabled an increase in speed to value of products and services to customers which is an area I have been a champion of for 20 years.

Where do you think the RPA and intelligent automation market is going? Any trends you've noticed?

There is a trend currently in ERP and CRM vendors adopting principles of RPA in their systems and AI in operations and analytics, however, these are proprietary and difficult to maintain and extend. Apart from consolidation in the market with niche players being purchased, the real revolution will be the commoditisation of the processes being built.

There is already a move in a couple of companies where RPA consultants are being compensated for sharing process or process snippets, and marketplaces will eventually be created for solutions that are executable by any RPA orchestrator and Robot.

How has and will intelligent automation affect us in work and at home?

Automation is already in the home and acceptance of smart connected devices has increased socially far quicker than it has in the workplace (for example, Ring, Nest, Alexa, TV, Smart whitegoods, Smart security and so on and so on). Adoption and expansion in the workplace will change the nature of work. However, the drive in AI and RPA is generally in white-collar professions which are catching up with blue-collar manufacturing where robotics has been in place and expanding over the last 50 years. Work will adapt around the robotic process and Ai, and while a shift to other activities requiring human intellect and physical capabilities will emerge, I suspect the impact will be similar to the advent of the production line, computers, email and the internet.

What challenges and barriers have you come across/overcome as a new entrant to the intelligent automation market?

The challenge as a new entrant is still education. Organisations understand the potential benefits of automation but do not understand the process, best practice or the realisable benefits that automation could have on their business both tangible and intangible. This is exaggerated further with AI where there is still great confusion, drive, partly by the media, but more by a lack of understanding of data and computer science and the practical business implications by CIOs and CTOs.

Bot Economy

This massive growth of RPA is due to the fact it is much faster to implement and has a much lower price tag than an IT overhaul. Businesses can see a return of investment within months after deployment, as opposed to IT changes which can take years to implement, let alone to see a return. However, in real terms, RPA still requires a hefty initial investment which could start at about £100,000 for infrastructure and licensing of several bots. Furthermore, dependent on a process's complexity, RPA development can take 3 – 6 months to build, and RPA developers don't come cheap (starting at £500 a day). With the high RPA developer salaries and the ever increasing demand for speed, business units are unsatisfied with the speed of many RPA initiatives.

This has caused Centre of Excellence development teams and suppliers to seek ways to increase the speed of development. This has given rise to the 'Bot Economy', where technology suppliers and RPA vendors sell or give away for free pre-built RPA bot components. These are standardized subprocesses or snippets of RPA code, such as logging into SAP or carry out a Know-Your-Customer process that can be downloaded so a developer can plug it into their existing RPA process.

This saves developers a little time by not having to build that part. These 'mini-bots' or code packages may save a day here and a few hours there, but these savings can quickly add up and exponentially outweigh their costs as developers continue to reuse these 'mini-bots'. Though businesses like to see themselves as unique, those who are in the same industry or market sector have many of same processes and applications, and the ways they are used are the same.

For all these processes it makes sense to acquire a handful of these standardized code snippets, such has how to add someone to a CRM system such as Salesforce or to download information from a financial website like Bloomberg. Furthermore, these mini-bots can provide plug-and-play Ai, where a mini-bot can be purchased, downloaded and integrated and immediately have capabilities such as image recognition, chatbot functionality or Machine Learning embedded into your process in minutes.

Blue Prism, UiPath and Automation Anywhere all have their own marketplaces or bot stores where technology companies from around the world can upload and sell their pre-built bots. This is great exposure for these new and established software providers, and great value for businesses new to automation.

Here are the bot stores of the three leading RPA vendors:

https://go.uipath.com/

https://digitalexchange.blueprism.com/dx/

https://www.automationanywhere.com/products/botstore

This however has raised concerns from potential buyers, such as the ownership of intellectual property, the level of support provided by the mini-bot vendors and the security aspects of downloading foreign code to be used inside a company's firewalls. Perhaps the rewards outweigh the risks in some instances, and some bot stores have taken steps to vet suppliers and provide some level of security checks of code added to their site, however the responsibility will inevitably lie with the business themselves.

The bot economy even extends to AI algorithms and Machine Learning solutions such as Amazon's AI algorithm marketplace, which is mainly for data scientists and AI engineers to find solutions to enhance their projects:

https://aws.amazon.com/marketplace/solutions/machinelearning/

With such a flood of new entrants into automation and AI due to fairly low barriers to market, the area of automation and AI has quickly become overwhelming. Business leaders are struggling to decide which intelligent automation products and platforms are most suitable for their business needs. Businesses consequently need to hire specialist RPA and AI analysts and consultants with knowledge of these markets. With this expertise, businesses can assess vendors and their products and get to a point where they can run POCs (proof of concepts) to see which solutions are the best fit and will work in practice. Some large corporations and consultancy firms have formed entire teams of analysts to do this vetting for themselves and their clients.

Working as an analyst I've trialed and reviewed many intelligent automation solutions, heard numerous pitches and spoken with many solutions providers.

GUEST: Tabscanner

*To understand the market through the eyes of a vendor, I got in touch with **CEO and founder of Tabscanner, Rashad Al-Safar**[3]. I was grateful to catch him and pick his brain before he jetted off to another continent. Tabscanner is an AI receipt-data-extraction technology accessed through a cloud-based application programme interface (API). The technology automates the capture, classification and validation of receipt image data and provides this data to a digital worker to enable expense automation.*

How does Tabscanner fit into the RPA/intelligent automation space?

---

[3] RJ Alsafar. https://tabscanner.com 2019

Our technology goes hand in hand with RPA eliminating the manual validation and data entry of paper receipt and invoice documents. We saw a perfect fit between our technology and a simple drag-and-drop tool that be utilised on an RPA platform. With this in mind, we were able to form a technology alliance partnership with Blue Prism and are now working with other RPA providers to enable further access to our API within the RPA space.

What trends have you noticed in the computer vision arena and their impact on automation? Where do you think this is going?

There is a rapid movement towards all forms of automation powered by AI and the advancements in computer vision technology. I see a future where data capture, recognition and classification will exist in virtually all fields and industries enabling people to pursue more creative tasks rather than repetitive ones.

How is or will AI and cloud technology impact computer vision?

AI will effectively power computer vision through intelligent learning. Cloud-based technologies allow for faster and more scalable solutions which allow greater flexibility and accessibility to computer vision technologies.

What are the biggest or most common struggles and frustrations companies are experiencing when trying to find and identify the right intelligent character recognition (ICR) solution in a crowded market that meets their unique needs?

I believe companies should do more research into potential solutions and emerging technologies, as the ICR industry is advancing rapidly. It is important to identify the best and fastest evolving technologies, and choose highly specialised services in order to grow and scale into the future. Finding the right solutions for each individual ICR requirement would be the biggest challenge and therefore, fast and efficient analysis of the options is vital to make the right choice.

The Impact on BPO

Labour Arbitrage

For decades, we in the West have outsourced many of our high volume, low-to-medium complexity processes overseas, such tasks as first line support call centre representatives, Level 1 IT support, and data entry and processing tasks. Companies saw that they could save up to 30–50% of the cost of an employee here in this country. Taking advantage of this salary arbitrage moved lower skilled jobs overseas, however just as with any arbitrage advantage, the gap has slowly started to disappear as more and more western companies did the same.

Many of the processes that had been outsourced overseas are the exact type of tasks fit for automation: high volume, low variability and rules based. In this new 'bot' world, the cost of a digital worker is 10% that of a full-time employee, and the digital worker provides greater accuracy, data integrity, and security. Consequently investment has started to move work away from countries like India, Malaysia and Philippines, and back onshore, bringing rise to a massive demand in automation developers in United Kingdom, other countries in Europe and the United States.

GUEST: Edge Tech

*I reached **Harrison Goode, Co-founder of Edge Tech** to hear about the market from the viewpoint of a **specialist recruitment firm in the UK**.*

Edge Tech Headhunters is an emerging technology headhunting and recruitment company based in the United Kingdom who operate globally. They recruit top talent within emerging technologies including RPA (Robotic Process Automation), IoT (Internet of Things), AI (Artificial intelligence), Data Science/Machine Learning and Blockchain. They pride themselves on providing a comprehensive network of highly skilled and niche professionals globally coupled with their expertise of being able to headhunt and place the most in-demand skills and professionals into exciting roles and companies.

What struggles and frustrations have your clients experienced whilst trying to implement automation/Ai?

The main struggles our clients experience is the lack of talent and the war for talent. Without us, our clients struggle to secure the best professionals in the space so they can successfully implement and complete their RPA/Ai programmes of work. That is where we step in to help and enable them to be successful.

Where do you think the RPA and intelligent automation market is going? And how has or might it affect us in the future?

I feel that the RPA and Intelligent Automation market is moving more and more toward AI applications and will soon sit under one AI or Transformation umbrella as one part of a company's strategy. We will see this make its way down to smaller businesses, and in the future, it will simply be a part of everyone's working life without even having to think about it.

The Big Threat to BPO

Intelligent Automation has threatened the BPO (business process outsourcing) industry, and no country is a bigger player in the BPO than India due to its lower employment costs and enormous talent pool of professionals. We've seen many companies in India need to adapt quickly to provide digital BPO and enhance their technical capabilities, whereas other companies who have been a little more nostalgic about the old ways are at risk of being left behind.

With the labour cost of RPA increasing in countries like the United Kingdom and the United States, overseas BPO business leaders have come back with yet another attractive proposition – for companies to outsource their RPA development to India, where over the last few decades the workforce has been improving their technical skills. Though senior RPA developers/engineers are currently onshore (or perhaps nearshore, out of the cities where labour and facilities are cheaper), many companies now have large RPA support teams overseas to provide out-of-hour or even 24-hour level 1 support.

Outsourcing overseas is particularly attractive to those companies who have scaled their digital workforce to multiple bots or RPA-as-a-Service (RaaS) firms who manage or support bots for their clients. Because many bots will run 24 hours, they will need on-call support to ensure their continued running (though it is generally assumed that one support engineer can manage up to 10 bots).

4 Choose the Right Solution

Choosing the right RPA solution for your needs is an important decision, because not all platforms are alike. Furthermore RPA many not be the right solution for what you are trying to achieve, as there may be an over-the-counter solution that's better suited to your needs for many reasons. It may be cheaper and much quicker to implement. It may have been built specifically for your process, for example, for automated marketing platforms.

RPA is not plug-and-play, and it requires a fair amount of time, resource and investment to get things going. But let's assume you've done the due diligence, and realise that RPA is the best course of action. As previously mentioned. there is a myriad of more than 50 RPA suppliers with the three most popular being UiPath, Blue Prism and Automation Anywhere.

Once you've decided on the vendor of choice, you have a few options for working with them. Some businesses work directly with the RPA software vendor to build and support bots; however, most businesses use a third-party service provider or consultancy. The ultimate goal is to build a Centre of Excellence team in-house to build and support your bots.

Here are 7 tips when looking for an RPA platform provider and consultancy firm:

1) You need providers who will response quickly to change. These changes could be regulatory changes, business practices, or IT changes (which can break bot processes completely).
2) Find solutions that don't require much programming, so that it's quicker to upskill staff.
3) Solutions should include real-time monitoring; remote, configuration, scheduling and API capabilities; and a user-friendly dashboard to easily manage your virtual workforce
4) Bots should be application agnostic. They must be able to work seamlessly on desktops, Citrix, web-apps and your legacy systems.

5) To get the most out of RPA, your solution must be scalable, so you can flex your workforce to meet seasonal ups and downs, and be ready to handle forecasted growth.
6) To ensure RPA stays cost effective, your virtual workforce must be able to handle a variety of different business processes, from back office to customer engagements (e.g., sending out letters to customers).
7) Finally, find out what the art of the possible is for the future. Is your chosen supplier looking into AI, cloud, NLP (natural language processing), or OCR (optical character recognition)? What else would you be able to plug into your RPA workforce? If you choose the right supplier, the opportunities could be endless.

Business Process Management tools

"If you can't measure something, you can't understand it. If you can't understand it, you can't control it. If you can't control it, you

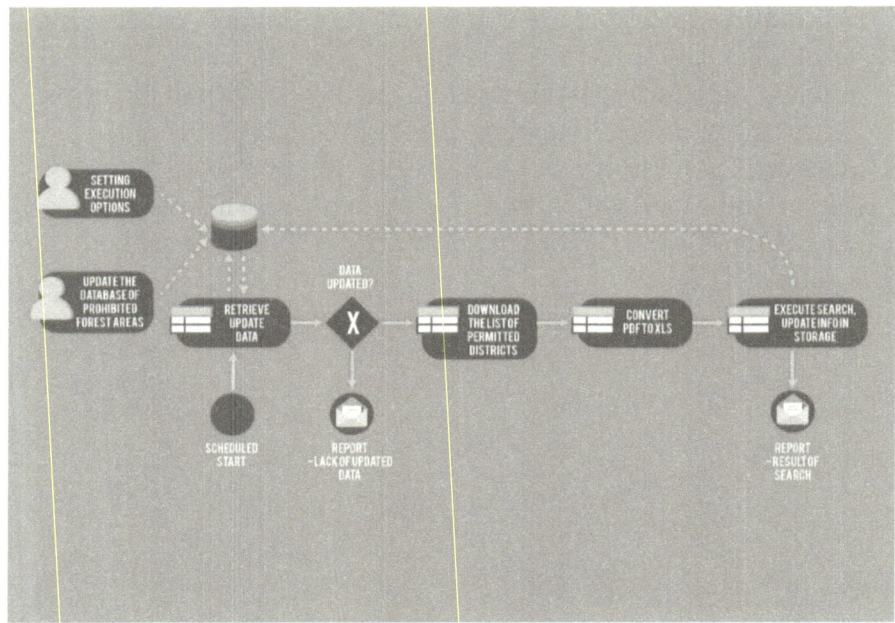

can't improve it" – H. James Harrington

BPM (which used to be known as business process mapping) helped a business understand it's inner workings, so that business analysts could better understand how BAU operations could be improved to enhance customer service and the quality of service or of product, all with greater consistency.

Now BPM has advanced to stand for Business Process Management, a methodology similar to Lean Six Sigma, which helps businesses make operations more cost effective. BPM is about performance improvement and automation, and it has many software solutions available. Lean Six Sigma is a more analytical approach focused on quality, and it identifies wastes to improve efficiency and performance. Combining the two is a powerful tool for digital transformation.

BPM software (BPMS) can tie several processes together in a department or an entire company into an end-to-end business process map and use elements such as workflow engines, forms, and automation tools (such as macros and RPA bots) to create a streamlined automated process. This helps form a 'digital nervous system'. In the last year iBPMS has emerged which adds intelligence to BPMS.

BPM software can collect data and metrics flowing through its processes and through the business, and it allows businesses to model different scenarios. Coupled with a data analytics tool, BPMS can provide powerful insights to business leaders as to what is happening inside their organization or what could happen, where their wastes and weaknesses are, and where there are opportunities to expand. Then a Lean analyst can apply Lean Six Sigma techniques to re-engineer the architecture to make those improvements

As your team learns more about Business Process Management, you will be able to see where RPA fits in, and how RPA, AI and other automation tools can be integrated into your business's operations, instead of being stand-alone tools.

5 The Evangelist

*"There's only one rule in this world, a small question that drives all success." ... "**What's in it for me?**" – Ray Liotta in* Revolver

Dr Stephen Covey says in his book *The 7 Habits of Highly Effective People*, first seek to understand before seeking to be understood.

The 'RPA or IA evangelist' plays a key role in ensuring everyone from senior leadership down to the staff whose jobs will inevitably be affected by automation are aware of what RPA is, how it can affect them and how it can benefit them in their specific position. To be successful they must be supported by senior management and the RPA/IA sponsor.

It's a different story for each level but the message and purpose is the same, to iron out any misconceptions so that everyone is on the same page, and to show the art of the possible and how this will improve their work life and objectives individually and collectively.

Conveying the message means giving presentations to different teams and holding lunch-and-learns, so teams and managers can see how RPA/IA is being used in the industry as a whole. Then teams can understand that it's not just something being enforced on them from above, but rather it's the new way the world is working. Through the clear congruent message from the evangelist, teams can hear success stories of how other companies and competitors are using and benefiting from this technology. Teams can watch demos of how bots are built and how they work – thus demystifying this mysterious technology and debunking myths and fears they many may have

Investing in a good communication plan to deliver to operations teams and difficult senior managers is vital as they will help implement automation and with be the ones who use it. Many RPA projects fail because of a lack of communication, with resistance coming from those who feel their jobs are at risk, and from a management who may feel they are being audited.

At a senior management level many managers and directors would have learned various information from different sources such as conferences or publications, and undoubtably they will have different ideas of what RPA is and can do. For this reason, it is important to ensure that the organization is spreading one consistent message. If not, confusion and frustration will certainly arise as some may expect RPA to be a silver bullet and have unrealistic expectations. This misconception will only escalate as its filtered downward, and it could easily turn into chaos as leaders create knee-jerk reactions to either scrabble for scarce resources or cut budgets all together after not getting expected results. The Evangelist is the key to avoiding this chaotic result.

I invited **Guy Kirkwood, chief evangelist of UiPath to share his experiences and thoughts**

Guy, I appreciate you sharing your insights, please tell us a bit about UiPath and what its unique selling proposition (USP)/value proposition is to the intelligent automation industry.

UiPath is the fastest-growing enterprise software business in history. Why? Because it offers a robotic process automation (RPA) platform that enables office workers to stop doing the boring, mundane and repetitive work that everyone hates and helps people to do the work that they want to do. This increases employee happiness and engagement, delivers time (arguably the most valuable resource) back to the business and enables companies to focus on what really matters: customer experience and intimacy. Bill Gates in the early eighties said that he envisioned a time when every household and every office desk would have a computer. Today, we believe that every person will have their own software robot.

What attracted you personally to the RPA/intelligent automation industry?

Having spent 20 years involved in the outsourcing market, specifically in BPO, the team I worked with in 2014 used the software which would become the UiPath RPA platform to automate a deal with a global technology company. The results were so staggering, I had a 'Road to Damascus' moment and realised that RPA was the future of work. In 2015, I was invited to join UiPath as employee number 28 and COO to help grow the business.

What trends have you noticed in the RPA and intelligent automation market? Where do you think this is going?

The main way in which automation and RPA is developing is in the ratio between unattended and attended robots; last year 64% of our licenses bought by customers were unattended, and today that figure is 46%. This reversal has come about as more organisations use attended robots to help their employees generate more positive and faster outcomes.

What we're also seeing is that AI and RPA are being combined to create truly transformational capabilities in four areas: visual understanding - viewing a screen in the same way that a human does; document understanding - any document in any format, including handwriting; process understanding - to identify which processes should be automated and self-healing when the process changes; and conversational understanding - so that robots can understand and react appropriately to the human voice.

Ultimately, we expect RPA and AI to disappear; not because it won't be used, but because it will be used everywhere.

How is or will AI and cloud impact automation markets such as RPA and BPMS?

The time of the monolithic service provider has gone. What we're seeing is the rise of the ecosystem where best-of-breed vendors in RPA, BPM, ERP, CRM and so on are working together to provide an integrated platform for business. This has been described by one of our competitors as a Venn diagram where the circles are not coming together in one 'blob' in the middle, but where they overlap very slightly and RPA and AI act as the glue between these systems. As an example, Oracle has worked out that it is much faster, cheaper and easier to do the final integration using RPA with the plethora of differing systems in their customers, rather than using the more traditional API route.

What are the biggest or most common struggles and frustrations companies are experiencing when trying to implement or scale intelligent automation?

Scaling is the number one concern of businesses using RPA and AI. Everest Group produced a step-by-step guide on how to scale from pilot to Pinnacle (where automation is spread throughout the enterprise). I've also written about how to scale in simple terms.

6 Run a Pilot: POC or POV

You've probably sat through multiple presentations from solutions providers and had to ask the same questions time and time again to tediously vet and shortlist them to find which vendor in theory would be a good fit. However only by trialing their software for real cases can you see whether they can deliver on their promise. POC or proof of concept is a phrase you've surely heard many times. The vendor comes into your company and runs a small-scale project on one or two use cases to demonstrate how their capability can provide real value and discuss how the software could be implemented on a larger scale. The proof of concept is the ultimate test if the solution will be fit for purpose.

At a small cost, compared to the full-scale version, the small-scale project is definitely a great way to see first-hand how a vendor's technology can work on your team's processes, and start to quantify the value it brings, and it can help you more accurately predict how fast you will see a return in an investment with them. As you go through the motions with the vendor, you can start to understand what is needed for the process to work. For example, do you need to improve the quality of your data with better data validation on the front end, does your process need to be redesigned, and so on.

Once the POC has evidenced its viability, it tells a great story that can get key stakeholders excited about the technology and get momentum going in your organization

One thing to note, however, is that you should be aiming for POV (proof of value). Having the new tech working on your systems and applications, and ingesting your data well is one thing, but it is important to evidence tangible benefits from the pilot. Measuring the process metrics (error %, handling time, flexibility) before and after can allow you to demonstrate POV, and it makes a much stronger message when presenting this back to the business and showing something tangible can remove a lot of doubts.

Are YOU Aware?

In this chapter we've looked at what RPA is with a little background on its history and how the market and technology has evolved, as well as where it is going. We've demystified a few common myths, explored the potential challenges and pitfalls as well as the benefits that automation can bring.

I hope that you see the importance of winning hearts and minds by being visible and taking stakeholders along with you through the transformation journey.

Now let's recap on the key parts of this chapter to look at how aware your core stakeholders are:

- Can you and your RPA team clearly articulate in a sentence or two what RPA and intelligent automation is?

- Do senior managers in Operations, IT and HR understand how RPA and intelligent automation can impact and benefit their teams and departments?

- Do directors and C-levels understand how RPA and intelligent automation can fit into the organisation's strategy? And do they have congruent realistic expectations?

- Do team leaders and staff members understand how RPA and intelligent automation can enhance their performance, productivity and job satisfaction?

- Do your RPA and Intelligent Automation Centre of Excellence and Operational Excellence teams understand the various myths and challenges as mentioned, and are they aware of how to avoid common pitfalls?

If you've done a good job communicating, everyone is now aware of this powerful, shiny new tool. They are bound to be very excited about it as they begin to think of how they can apply it to achieve their individual goals. However, overexcitement without control can cause things to go a bit haywire. The next challenge is a lot like herding cats…

Lots of ideas …in different directions

A team without a clear direction will get nowhere fast. Many times teams without clear goals or aligned objectives get flooded by countering opinion and end up draining scarce resourcing in inefficient ways and burning money until either their initiative is closed down, or they are left to plod along in the corner, as people give up on realising the benefits they had once looked forward to.

Align your vision

"Start with Why"– Simon Sinek

The first step to doing anything new is to start with Why. Why is there a need for this change, where will we be after this change, and what new capability will we deliver? Getting buy-in and having a *shared* vision from the top is the first step.

The vision and the purpose of your CoE points your team in the right direction and tells your stakeholders where you're heading, with your strategy and objectives depicting how you plan to achieve this. Everything your team does is pinned onto these pillars, from the products you offer, down to the training, attitudes, and behaviors of your staff.

Without a clear vision, you won't know what teams or departments to enter, what services to provide, and what technologies to seek out. I've witnessed first-hand the aftermath when a team doesn't have a clear trajectory or focus. This is especially true in RPA and intelligent automation. With such powerful tools, if the people at the top do not have an aligned vision of how to use them, chaos will occur at lower levels, with multiple teams using different platforms, and vendors having contrasting approaches on how to assess opportunities or implement automation.

It's highly inefficient to have competing RPA teams in the same organisation using different methodologies or best practices. If departments were completely isolated, it probably wouldn't be as big an issue. However, end-to-end business processes run across different teams and departments, and inevitably the bots will need to interact with each other. For example, if in your organisation RPA team A builds a bot with poor error handling and this data is sent to another department who have a bot built by team B with different coding standards, then it is likely that this will become messy before long. In addition, if the team that builds automation doesn't follow the same standards as those who support it, maintaining the bots would be difficult. Furthermore, if developers aren't aligned with IT or compliance, the business could become vulnerable to security or data breaches.

Intelligent automation shouldn't be treated any differently than any other new capability being introduced to your business. You should have a Centre of Excellence in place to ensure consistency across programmes and projects and ensure that these activities align to the overall strategy. CoEs give assurance of quality and that the right people are making the right decisions based on the right data.

However, in reality your company is most likely siloed, with different senior managers having contradictory agendas, and if there's no support and sponsorship from the top, departments will fight over where the CoE should sit and most likely will set up their own 'CoE teams'.

Silos (Department v department, the business v IT)

The Automation Centre of Excellence is the bridge between Operations and IT; there can potentially be disagreement on where it should be managed. However, some may feel that, as this is a digital transformation capability, it should sit with the Change team, or perhaps with the digital solutions team.

Wherever it sits shouldn't much matter, as it should be given the freedom to operate independently and as a connection between Operations, IT, and Change. If your organization is big enough to have these departments, the CoE should be formed from members of these three departments to bring the experience and knowledge of the processes from the Operations team; skills in business analysis and project management from the Change team; and IT development, support, and infrastructure knowledge from IT. It's vital to bring external skills from industry experts to ensure successful execution; however, you should look to upskill internal resource from these three teams.

When starting out, your CoE should be a *centre* of excellence, singular. So it's critical that automation teams and rogue 'centres of excellence' don't start popping up throughout your organisation, as they will surely not be a *'centre'* of anything, and without aligned sharing of lessons and expertise, *'excellence'* will not be seen.

As with any new change initiative, there needs to be a central contact point, and congruent message needs to go out corporation wide. To continuously improve on implementing new changes, good knowledge management and a collection of lessons learnt will assist the team in 'failing fast', that is to learn quickly, to progress in executing faster, smarter, and with less hiccups as your team matures. If several independent automation teams exist in your company, these individual teams will learn slower. This will also confuse the wider population, as they may have different approaches, and at a senior level comparing value of opportunities may be like comparing apples to pears.

Involve IT early

As previously mentioned, RPA was gifted to 'the business' as a solution to remove technical roadblocks in a matter of months and at a lower cost. Customers continually want faster, cheaper services, so operations departments need to be more agile, as IT's cumbersome system changes can take years and cost far more than a few RPA robots. RPA has also given operations teams a lot more control on what to change, and they can make continuous improvements and implement modifications at speed.

However, the one thing many organisations fail to realise when bringing RPA into their organization is how vital the IT department is when it's time to scale, and so they don't get IT involved early enough, which causes issues and delays later on. I've heard of many companies where various back office teams have built their own assisted robots (which sit on the users' computers) to run on their legacy systems. However, as they had not engaged with IT, they were not made aware ahead of time of user interface changes or system downtimes, which caused their robots to fall over. Your Centre of Excellence team, no matter how small, needs to have open communication with IT so you are aware of roll outs of new system updates. This is also important to ensure that test environments have the latest versions of applications.

Furthermore, when you need to scale and have unattended bots working in the background, the bot will still need to sit somewhere, so IT will need to be involved to provide server space to host and maintain the virtual machines and virtual desktops that the bots will operate on.

7 Centre(s) of Excellence

Setting up

When setting up a Centre of Excellence team, there are several things to consider. However, the first is whether to set up the CoE capabilities in-house, outsource them, or do some sort of hybrid.

You may want to hire a team of experts to come in, set up your CoE, and gradually over the next few months roll off certain roles to in-house staff as they gain training and experience.

Whether you build the team internally or you outsource, you will still need a team of experts to come in and lay out the **framework and the underlying processes and tools** to set you up for the future:

- Governance:
    - Governance model and policies
    - Risk assessment and reporting tools and KPIs
    - Role profiles and team structure
    - Service definitions and catalogue of what your CoE will provide to the wider organization
    - Templates, forms, and controls
    - Capacity and capability management of the team
- Design team
    - Lunch & Learn Evangelist pack for awareness and training for Automation Champions
    - Opportunity identification and assessment tools
    - Automation catalogue and prioritization tools
    - Cost models and cost benefits analysis (CBA) tools
    - Development standards and deployment processes

- Support team

  o Support model

  o Change management for releasing new bots or bot modifications

  o Hyper-care criteria to ensure quality bots are rolled out

In this image you can see how different roles in the lifecycle can

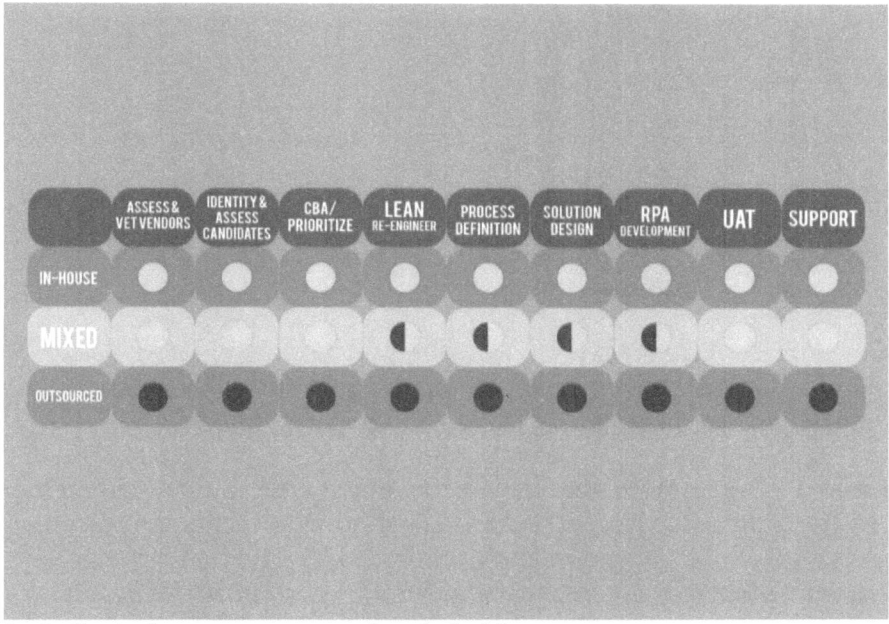

look for in-house, outsourced or hybrid teams

Analysis tasks are **easiest to teach and pass on knowledge**. You can hire an RPA or Lean expert to come in with their toolkit and train your staff to do some of the following:

- Solution assessment and vetting

- Process identification and assessment

- Cost-benefits analysis

- User acceptance testing

Tasks which will require **longer-term mentoring and training,** where you have a mixture of hired experts and permanent staff:

- Lean thinking and process re-engineering (including root cause analysis (RCA)), process definition, and solution design

- Building the solution

- Supporting the solution

How mature is your CoE

I have seen several levels of maturity. Which one are you?

Level 0 (Explore stage): You've just started looking into RPA and automation; you've not deployed anything. You've seen a few demos but have not yet decided which vendor(s) to use.

Level 1 (Experiment stage): Your team have self-built automation robots using a free license to understand how the technology works and/or you've engaged with preferred vendors who may have given you some licences for free. No real structured approach exists for assessing or implementing opportunities.

Alternatively, if you have outsourced your CoE completely, then your maturity would also be Level 1, as you have not yet developed any in-house capabilities.

Level 2 (Pilot stage): You've hired a hybrid RPA developer who also does the business analysis work or a developer-BA duo, to find opportunities to automate, and you've deployed a POC (proof of concept). Some RPA documentation and assessment tools exist, and you have a logical approach to assessing opportunities.

Level 3 (Team formation stage): You have a few expert team members doing some of the roles that make up a CoE team but do not have a complete structure; perhaps you have a solutions architect, developers, support engineers, RPA analysts, PMs, and a Lean Analyst. You have documented processes, a framework, and a toolkit to generate consistent automation solutions.

Level 4 (RPA factory): You have a polished team of experienced professionals who have delivered automation in several environments, as well as automation champions throughout the organization as advocates. A steady pipeline and a refined framework exists to churn out automated solutions at a consistent quality and speed. You have a complete toolkit with controls and reporting tools.

Level 5 (CoE as a Service): Your experienced team can provide their services to internal and external clients. You have clear terms of service, SLAs, and governance processes.

If you have a large organization, you have probably set up several satellite RPA teams that report to you.

I asked previously whether your company was an *early adopter* (do you already have a Centre of Excellence team and have you learnt from the implementation of several bots), or are you a *laggard*, last to find out, last to move, and the least to benefit from new technology.

If you're a laggard in RPA, then you can appreciate that your business's future is at risk and you're probably being surpassed by faster, scalable, more agile competitors. You may have started to hemorrhage customers and are losing money trying to keep up. Again, if you're in financial services, including banking and insurance, or you're in health care or utilities, these markets were first to start embracing this technology, so you being a laggard in these industries leaves you in a dire state.

If this *is* you, do not fear, in this section you can discover how you can catch up and overtake, as many competitors who are ahead of you are most likely battling with many of the internal conflicts and pitfalls discussed in the previous sections because they didn't get the foundations set up correctly from the outset.

If you are the early adopter who's now putting out fires, trying to resolve conflicts, and struggling to scale, then this will enable you to get traction once again. So, let's look at setting up an organized Center of Excellence team, with a robust implementation framework.

Structure

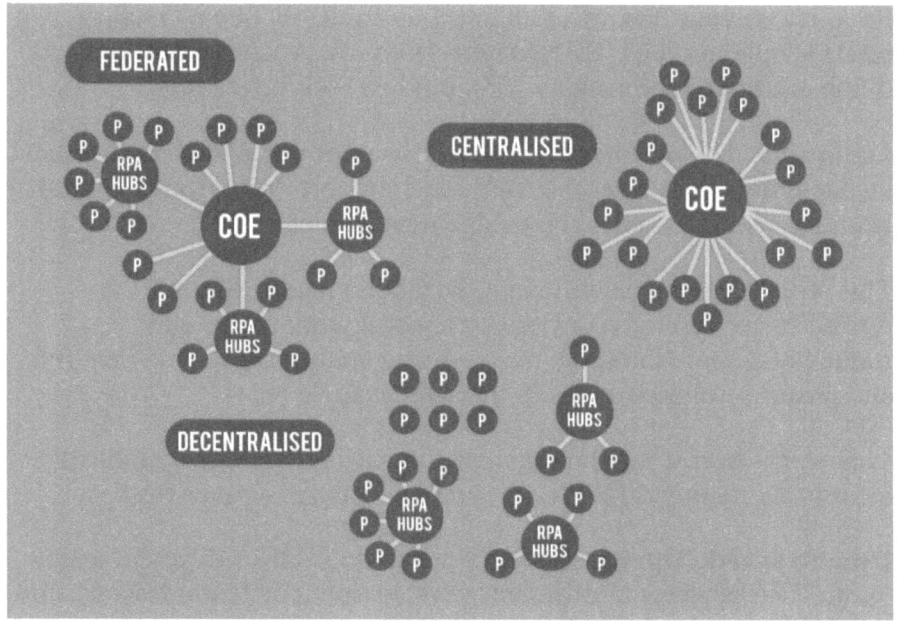

There are two main types of CoEs, centralised or federated. However, the maturity of your CoE team and the geographical location of your operations teams may determine which structure is right for you.

Centralized: This is where you generally should start. You have one team that manages all RPA opportunities happening throughout the organization. This is good for companies where the CoE staff sit together and all staff are based in the same location, or the teams and business units can be easily reached.

One centre exists for knowledge, learning, resources, governance, and tools. This way, instead of having two teams learning different lessons, you have one team learning twice as fast. Even if you start doing it wrong, you will 'fail fast', refine, and perfect one framework. If the entire company is using the same RPA vendor, you'll also benefit from economies of scale in license costs

The other benefit is that you will have a consistent message about RPA across your business that gets buy-in. RPA, like all change, can make people nervous (RPA maybe more so). A lack of buy-in is one of the top three reasons why RPA (and AI) initiatives fail

Federated: This is where you have resources embedded in different business units, however they still feed into and draw from a central team for training, tools, knowledge management, and governance.

This still requires a central team, so that no matter where RPA resources are deployed or based, they still work to the same framework, share the same message and analysis costs and benefits, and prioritise opportunities in the same way.

This works best when the organisation has different geographical locations, creating difficulty in sending resources out centrally.

Decentralised: The worst thing you can do is allow different teams to run RPA projects in different ways, learning different lessons, and using different RPA vendors. Aside from this being an extremely inefficient use of money and resources, it will inevitably get very messy and create unnecessary politics, and you will eventually stall your digital transformation strategy.

Roles and responsibilities

Now let's take a look at the roles and responsibilities in the Design

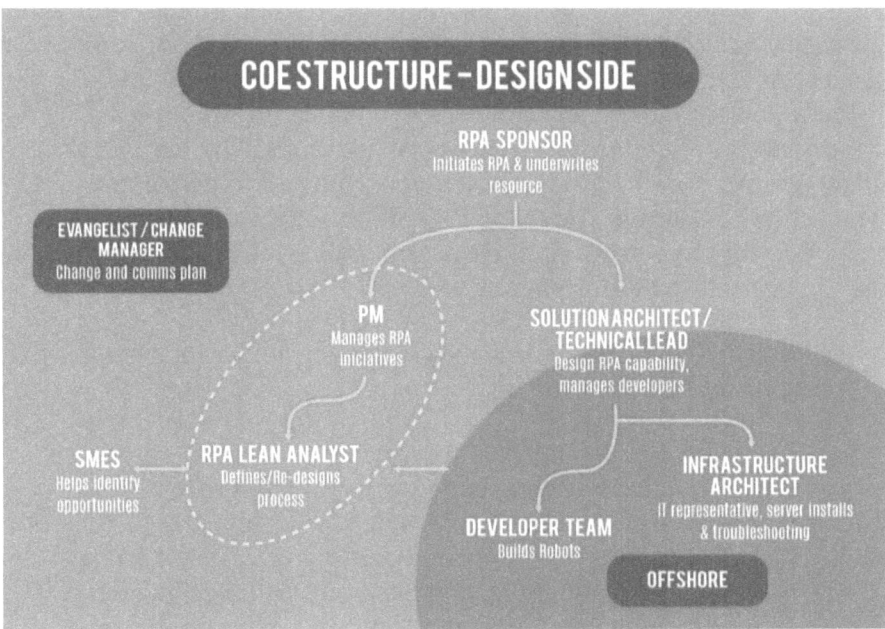

side of an RPA Centre of Excellence team. This will help you highlight which skills you may be missing.

The RPA sponsor: If you're the senior level management who brought the new capability into the organization and control the budget that funds this team, then this is you. Perhaps you were advised or recommended to use RPA by a subordinate; however, you will be accountable for its successful execution. To fully support the team, its's up to you to pitch this to your senior peers, remove roadblocks, and encourage cooperation amongst departments to aid a smooth rollout.

As part of the senior leadership team, along with the operations directors and the chief operations officer (depending on your company size), you need to understand how best to invest your pot of money to reap the biggest benefits. The programme manager with the assistance of the senior analyst(s) will be able to show you the process landscape of your organization, to clearly and plainly illustrate where the biggest and easiest opportunities lie, your optimisation options from a plethora of solutions. This will leave it up to you to pick and choose which opportunities to go for and which solutions to use (be that RPA, off-the-shelf automation, AI, Lean Six Sigma, outsourcing, or a combination) that will generate the highest return on investment.

Later in the book we show how the CoE team can do the analysis to present this picture back to you.

Your CoE team may be an internal service provider where you provide RPA development and support services to process owners throughout your organization. In this setup, your internal customers are coming to you, or perhaps you are deploying business analysts (BAs) to them to identify opportunities. Your team will need to vet these requests, and you will need a mechanism to determine which opportunities to pursue so that your CoE provides the highest value back to your cost centre, whilst keeping clients happy so they keep coming back and your pipeline stays full.

The evangelist or change manager: This may not be a full-time role in a small CoE team, although it is a very necessary one. The change manager most likely would take on this role; however, it could be the programme manager or even a senior BA, jointly or solely. This is the person who sells the capability and technology benefits to department heads, managers, and end users in order to get buy-in throughout the organization. They run workshops, organize lunch and learns, and bring in vendors to give live demos so department leads start knocking on your door with potential processes that they want to automate.

The evangelist is responsible for creating and/or delivering the communication plan to keep the CoE visible in the eyes of the business, keeping teams engaged and onside with the eminent changes. The evangelist is attentive to the concerns of stakeholders and squashes any myths or misconceptions, but also builds up a sense of urgency and thus builds demand for automation, so that the path is clear for the BAs to come in and work with these welcoming operations teams.

The programme manager: You are responsible for delivering this capability. You oversee and manage the progress of various RPA projects that are going on, and so you must have visibility of progress, blockers, and dependencies. This may include resource management, as you should have a clear view of what is in the pipeline. A larger team may require a Programme Management Officer (PMO) to keep track of each project's progress, and having checkpoints gives you added visibility and control. To keep projects moving, remove excess approvals by setting tolerances (time, quality, budget, scope, risks), allowing you to manage by exception. You'll have several business analysts/project managers providing status updates, and this is all rolled up to present back to the steering group/leadership team which is led by the sponsor.

The programme manager and the business analyst(s) form a partnership to liaise between the business and RPA Development team at the different seniority levels.

The RPA business analyst/project manager: This is the central role in the RPA lifecycle as these individuals can be involved at each stage, from vendor selection, identifying the problem, designing and building the solution, to ensuring that the bots are being supported correctly. They most likely will also reevaluate existing bots for continuous improvement to enhance bots as processes change or as new technologies such as AI are introduced.

In smaller CoEs, many roles can be performed by the same person. The business analyst is the most versatile and so should have the right amount of knowledge to regularly wear various hats, from evangelist, to project manager, and in some situations hybrid analysts have the technical skills to develop bots too. In many cases, the BA takes on the role of project manager. The BAs gather requirements and assess opportunities, and facilitate root cause analysis and solution design workshops, but they may regularly be expected to manage the RPA project through to implementation, and beyond.

Biased as it sounds from the viewpoint of a senior BA/PM for many years, this is the most important and essential role because you're responsible for turning the business problem into a technical solution. You're deployed into teams as the intermediary between the business and the RPA/tech team as translator for these two very different worlds. The business generally won't fully understand the technology, so the BA is there to explain how it works in non-technical terms to help operations envisage how it can work for them. Furthermore, the RPA developers won't have any understanding of the processes and may not have seen or used the applications before, so the BA needs to explain this in detail sufficiently that the developer knows exactly what the bot needs to do.

Some small start-up CoEs seek out hybrid BAs that can develop bots to get two for the price of one. However, this might not be scalable. In many situations, it could cause bottlenecks and, if the solution designer is also developing the bot, it is likely that they may make assumptive steps themselves either in the design or in commenting the code, and their knowledge may not be fully captured if they left, for the next person who takes over that bot. In addition, this hybrid set up allows the BA-developer to 'mark their own homework' which encourages laxity. As long as you as a BA have a good knowledge of the platform and underlying code to be able to build a simple bot yourself, then you will be able to communicate with developers.

Depending on how RPA was introduced into your organization, the sponsor may already have agreements with a specific vendor to start running POCs. Otherwise, the sponsor may require the BAs to analyse vendors, looking at pros and cons to find which would best fit their business objectives. So the BA is called upon right at the start. We will look in the following chapters at how they bring value throughout the whole lifecycle.

A final point to note, though rare as we are, RPA BAs who are Lean Six Sigma qualified can help any team get the biggest bang for their buck, as they are skilled in removing waste and streamlining a process first, even before automating it. This way you may require fewer bots, as opposed to automating the process as it comes in a Heath Robinson sort of way.

The RPA developer: RPA vendors have created very easy Drag & Drop user interfaces on their platforms, and with free versions that can be downloaded onto a computer, essentially anyone can build a basic RPA bot in under an hour. This isn't advisable, however.

Developers have experience of how RPA works in various scenarios and can avoid pitfalls of which a novice might not be aware. Hiring an experienced developer will ensure bots are developed fast, using best coding practices, and will also help reduce the amount of defects found in the bot, which may only become apparent once it has been running for some time. Taking the risk of launching poorly built bots will end up costing you in the long run in support costs, or in having to re-build it in the future, not including the cost of the bot processing data incorrectly.

 Your developer(s) will form a partnership with the BA. They work very closely together to define what the technical solution needs to look like using the process definition document (PDD) which the developer then converts into the solution design document (SDD). The developer and BA will also work closely during the user acceptance testing (UAT) phase, as the BA will facilitate this with the end user to ensure that the bot handles different scenarios in the right way and that the end user is happy with the outputs.

Developers should all use the same best practices defined by the CoE, add to the central knowledge base their experiences and lessons they've learnt on previous projects, and help develop documentation needed to pass the robots over to the support team. Preferably, developers will be working on one platform. However, they may need to be clued up on a handful of RPA platforms that are used in the organization.

The solutions architect/technical lead: Whether or not the CoE has a few developers or just one, the CoE will need a technical lead to offer support and guidance and uphold best practices. The CoE will also need a technical solution designer. This individual will be responsible for designing the infrastructure needed to build, test, and support the robots (virtual machines/desktops, servers, etc.) as well as supporting training and capability management of the development team to ensure that all developers are building bots to the same standard and deploying bots and upgrades in the right way.

GUEST: Solutions Architect

I caught up with **expert Solutions Architect David Orton,** *who has over 7 years of intelligent automation experience, to get his **top 10** list of what a company new to RPA should be thinking about:*

1.   Senior management engagement with an expectation that time, effort, and money will be required to make RPA a success.

2.   Active process owners with a solid understanding of the processes, how each fits into the bigger picture of the organisation, and who can provide access to local subject matter experts who execute the process. These people will need to be prepared to invest time into RPA.

3.   Fully supportive IT department that understands RPA and is keen to make it work and are not threatened by it.

4.   Fully supportive security, architecture, and CTO teams that understand that RPA is not traditional development, are keen and see digital workers as similar to human workers, enabling access to systems without undue obstructiveness.

5.   A clear vision of what the organisation wants from RPA beyond simple, short-term cost saving and a view of how they wish to operate (fully independently, with a partner, or completely outsourced).

6.   A vendor and/or professional services partner that understands the goals and culture of the organisation. A clear vision of the organisation's aspirations and technical capability will make software selection easier.

7.   A dedicated RPA team, ideally drawn from internal teams with extensive process knowledge but technical aptitude to learn RPA, and supported by an experienced partner, particularly in the early days.

8.   A realistic view of early achievements, including the possibility of financial loss in the first year, but with a solid roadmap of processes.

9.   Willingness to start small and learn RPA on the simplest processes.

10.   A strong understanding of the organisation's technical infrastructure and applications, particularly where Citrix and non-web applications or java applications are used extensively.

Anything else from a technical standpoint to consider?

- *Virtual machines or desktops?*

- *Locked office or data centre?*

- *Develop on local machines and upload or develop on virtual machines?*

- *Will VMs live in the same security zone as other user PCs or in their own zone?*

- *Will VMs have their own credentials and how will they be managed?*

- *Will VMs have a standard desktop build or a bespoke build?*

- *How will developers access VMs? Remote desktop, Citrix, or something else?*

- *How will patches, security updates, backups, and virus scans be managed on VMs that may need to run 24/7?*

- *What level of access will developers have to make changes on VMs?*

- *Will VMs be governed by existing group policies or bespoke?*

- *General considerations of resilience such as multiple data centres, UPS (uninterruptible power supply), backups, and general disaster recovery.*

Automation Champions: These are staff who are not on the CoE team, but are members of staff in operations that have been trained by the CoE in Lean thinking and RPA at an intermediate level to conduct various tasks to support the analysts and developers in the Assess, Build, and Test stages. They are trained in how to use CoE tools and templates and understand how these fit into the wider implementation process.

They are extensions of the CoE and are advocates of the technology within their operations team. Being the eyes and ears of the operation, they can feed back to the central team any concerns or resistances they've observed

CoE structure and services

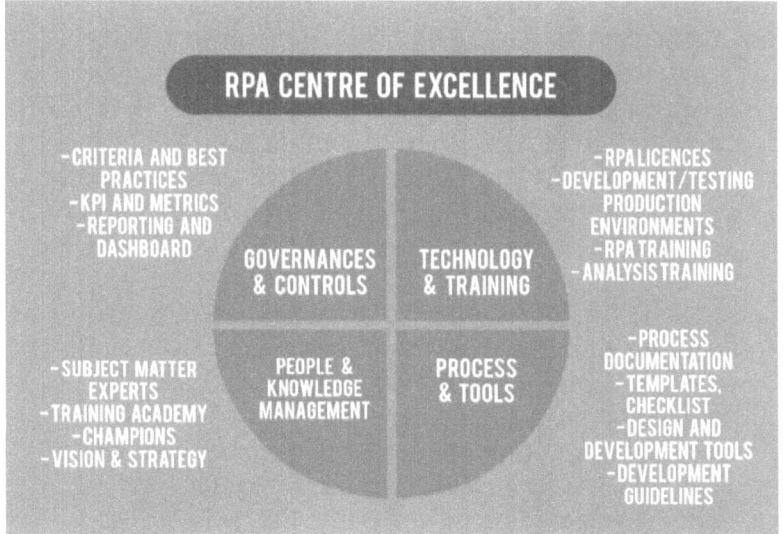

The four areas that CoE teams should provide as a service to the organisation are Technology & Training, Processes & Tools, People & Knowledge Management, and Governance & Controls.

Technology & Training

One way to ensure that no department or team decides to create a rogue robot is to centralise the access to the technology, both hardware and software. That said, some vendors provide attended bots and free versions, which is great for fast setup to trial the platform, but it does mean that anyone in your organisation can download the software and start building. This can be hard to control organisation-wide; however, some do believe that RPA will eventually become democratised like an Excel spreadsheet, where users build their own automation with Macros.

Aside from this, if you control access to servers, virtual machines, development and production environments, and licences for automation platforms, this is the best way to control how the technology is used and who can use it.

Once this control is in place, you can roll out training to ensure that those who will be using the technology will abide by your best practices and use the technology and tools correctly. Training includes not just technical training like RPA (e.g. how to use Automation Anywhere, Blue Prism, UiPath), SQL (for database management), and C# (which can extend its functionality. VB.NET and C# are the underlying languages for more RPA tools), but other skills needed to successfully implement bots, such as RPA focused business analysis, Lean thinking (e.g. yellow or green belt level Lean Six Sigma), and Agile project management (e.g. Prince2 or PMP).

In a centralised CoE structure, you can have assurance that everyone is receiving the same training, which will provide consistency in the outputs of the various automation projects. Furthermore, if all opportunities are being measured in the same way, then senior leadership can be confident they are comparing apples with apples when prioritising, and the support team can be confident that development coding standards are consistent with each bot they receive.

Processes & Tools

Going hand in hand with training is having a set of tools that the CoE rolls out to its team members and 'automation champions'. By giving staff a set of tools (like a complexity matrix or an automation catalogue) and a framework (like the steps in this AEIO YOU method), this is a way to embed controls in your implementation process so that all projects follow the exact same method, from the type of data that analysts collect and how they identify and assess opportunities, down to how developers write and comment their code. Moreover, when analysts present their findings back to the senior leadership team, the outputs of these tools consistently illustrate the opportunities, project savings, and progress in the same way, no matter what project it came from.

In addition to tools, you also want to include template documents (like business cases, PDDs, and SDDs), code review guidelines, and criteria checklists for each stage. Having a clear process showing when each tool or template needs to be used, and having controls and conditions built into the tools, gives the assurance that all the correct information is collected, no steps have been missed, and the right authorities have signed off on each stage. There should be a toolkit for Vetting & Assessment, Design & Build, and Support & Maintain stages. Later on, we will look closer at a few of the tools your team can use.

People & Knowledge Management

The CoE is responsible for deploying personnel onto each automation initiative, though 'automation champions' already reside in that team and have had training from the CoE academy on how to assist with identifying and assessing solutions, and user acceptance testing.

It could cause confusion if a team independently tries to design or implement solutions themselves or hires someone externally to do so. Worse yet, they were to start running demo workshops with vendors, spreading a different message to that of the CoE's. All external experts should be hired through the CoE, and these experts should add further value and enhance the CoE knowledgebase during their time in the organisation. Centralising the expertise and training means that as a whole the organization can learn faster, and lessons learnt and hurdles of previous projects can be shared to help refine the implementation process and modify tools to reflect this. Knowledge management is vital for your CoE's success as staff come and go. It's surprising how little thought many companies have put into preserving knowledge before key staff members move on.

Governance & Control

Having the toolkits and providing the training will ensure that your CoE team runs like clockwork, keeps churning out automation solutions, and maintains a full pipeline ... *but* that's as long as these measures are followed. Governance enforces these controls and provides management visibility when controls aren't being followed. This can be through the use of RACI charts (who's accountable and responsible for each task), escalation points (how to escalate if tasks aren't being completed properly or on time), and dashboards to see which gate an initiative is at and if it's on-target or behind, and what the blockers are.

The CoE is also responsible for risk management and compliance issues, ensuring that there is a business continuity and disaster recovery plan. However, this is handled on the support side of the CoE.

Eventually, you will build a robust framework for the whole life of the bot, from identifying, assessing, prioritising, designing, building, testing, supporting, and ultimately enhancing your automation. Embedding controls and checkpoints and monitoring your team's performance at each stage will have you well on your way to building an 'RPA factory'.

Are you Aligned?

In this chapter we talked about the detriment of not having alignment which starts at a senior stakeholder level, and how having several automation teams working independently could waste time and resources and result in confusion and inconsistency.

We looked at the roles and structure of a CoE and the different maturity levels. Answer these few questions to see how aligned your organisation is:

- Do all senior stakeholders agree on how to structure and utilise the CoE team?

- Is the strategy for RPA clear and is all communication concurrent with each other?

- Does your CoE team have close ties with the IT department, and does IT understand their role in RPA?

- How mature is your CoE and do you understand how to get to the next level?

- Is there a central location where BAU departments can go to request automation and information?

- Is there only **one** "centre of excellence" in your organisation?

Otherwise can you combine several RPA hubs or centralise resources, tools, and knowledge so you can learn and improve faster?

- Are all RPA opportunities identified, assessed, and measured in the same way?

- Do all developers and support engineers work to the same coding standards?

- Do you have a standard tool kit, including checkpoint criteria and templates for each stage of automation lifecycle?

At this stage, your stakeholders are aware of the technology; they understand the benefits at a high level, and everyone involved in automation seems to be singing from the same hymn-sheet, so to speak. Great job, you've herded the cats. The hard bit is over. Now it's time to build momentum and get everyone involved by empowering them with education.

E: Educate & Empower

"Transformation is a process, not an event" – John Kotter

Businesses need change, and change needs people. If you want to implement these changes fast, you're going to need the right expertise in the core team. Hire external experts to get things moving, and transfer knowledge to existing staff.

Furthermore, with change, the biggest resistor generally comes from the teams and managers who are most impacted by this change. Educating them on the technology not only encourages transparency and openness of your Centre of Excellence, but it can also empower operations teams to provide support.

Empower teams to look out for opportunities and create a 'Scout culture',[4] where everyone is on the lookout for improvements. It's a lot easier and faster if the end users are identifying opportunities for improvements rather than just the business analyst.

The most valuable resource of a business will always be its people

It's a lot more acceptable to go along with a change that you helped implement than to succumb to a change that has been forced onto you and created by someone else. Yes, the articles online correctly state that RPA will replace some jobs, but it will replace them with new versions, and new creative roles will emerge allowing people to advance their careers. In an interview, Jeff Bezos, the founder, CEO, and president of Amazon.com, mentioned that 'we are so unimaginative of what future jobs will look like, as 100 years ago the concept of a massage therapist [or dog psychiatrist] was unbelievable'.[5]

---

[4] John Kotter. Our Iceberg is Melting: Changing and Succeeding Under Any Conditions

Jobs are evolving all the time, and we can appreciate that we need to evolve to survive as a business. However, it would be detrimental to just see RPA and automation as a cost-cutting tool. As discussed at the start, there are many benefits aside from FTE savings that can put your business at an advantage: namely, better customer experience, agile workforce, and faster and more available services, to name a few.

I've seen and heard of companies that have implemented RPA to remove FTE only to recruit new FTE down the line after the business has grown (but with the added cost of hiring and training new staff, and the added risk of them not having the same level of experience). Instead, it might have been possible to re-deploy the existing staff to other faster-growing business areas and keep that knowledge in-house.

Your business's real value is your existing staff with their immense knowledge of your unique business processes and your market sector. RPA should be seen as an opportunity to upskill your staff, building Intelligent automation capabilities inhouse, bringing in experts who get things moving fast, with the goal to transfer their knowledge to your team.

8 Bring in the Experts

Most companies will attempt to set up RPA teams themselves, try to scale, fail or plod along, and then hire experts to tidy up the mess they've made. It's a lot harder to clean up a messy CoE team where people have become set in their ways and (ironically) reluctant to change.

5
https://archive.org/details/CSPAN_20180526_053500_Bush_Center_Leadership_Forum_-_Jeff_Bezos_Conversation

What would be better, once you've successfully completed a proof of concept (POC), evidenced the value the technology delivers, and committed a budget, would be to hire an expert early on to come in to help set up a Centre of Excellence team. This could be a team as small as three people: a business analyst/project manager, an experienced developer, and a solutions architect.

When you hire experts as temporary staff, it's important to have a plan for them to train your staff, transfer their knowledge, and leave you with tools and frameworks for your inhouse team to continue using after the experts have left. As an 'Automation Success Manager', if you will, my counterintuitive mindset when going into a new team is to set things up so that my role becomes redundant, and the team can function self-sufficiently—that's when I know I've done a good job.

An expert that you hire to come in and help develop your CoE team should provide you with three things:

1. **Experience** of what does and doesn't work in industry, guidance to navigate your team around pitfalls, and experienced staff to help you develop a model that is fit for your organization.

2. **Standard tool kit** of templates, processes, and criteria that they've developed which they can tailor to your industry and organization.

3. **Training and mentoring** for your internal CoE staff to learn the ropes. The hired expert shouldn't work independently but should train staff on the job and consistently transfer their knowledge so that your permanent staff are confident to keep delivering once your consultant has moved on.

> Hire a consultant who can help you build up your capabilities so that the work can be handed over to you and you can be self-reliant.

GUEST: Broadgate Consultancy

I reached out to **John Vincent and Richard Gale,** *who are* **partners at the IT consultancy firm Broadgate Consultants,** *to get their take on this:*

Broadgate Consultants work with clients to deliver business-focused solutions that are tailored to their individual requirements whilst meeting the financial budget constraints.

Their success is driven by the quality of their work, practical attitude, and the strong relationships they build with their clients. Their managing partners have a combined experience of over 30 years in business technology, together with a trusted network of carefully selected associates you can rely on to deliver great results every time.

What is your approach to RPA/intelligent automation?

We are enthusiastic supporters of the benefits of RPA and are keen to emphasise the human aspect and how the change can be implemented in a responsible way. We take time to think about the role of existing employees, retrain staff to work in the new RPA departments, and carefully steer both management and employees through the new ways of working. We see RPA as one tool in a whole suite of technology that will enable organisations to fulfil their business needs in a more efficient way.

Our core business is to improve our clients' business operations, making them more efficient, effective, and ultimately grow faster and become more profitable. A critical component of this is to improve the way they work across people, processes, and technology. We see RPA or 'process automation' as a key tool to assist. The majority of RPA's components have been available for a while now, but the new products such as Blue Prism, UIPath, and Automation Anywhere bring the separate tools together and provide a structured framework to support automating processes. Throw in the potential for machine learning and AI, and RPA presents a very powerful toolkit.

RPA is exciting technology that could really change the way we work, taking the boring stuff off the table, freeing up staff to accomplish more customer facing/beneficial tasks. We are interested in the challenges of managing the change regarding the human work force and the integration of robots into the team. We think, as RPA implementations become more widespread, that the idea of some kind of corporate/social responsibility will be something that all executives will be wise to consider.

What struggles/frustrations have you experienced whilst trying to implement automation?

Lack of enterprise-wide approach, often a head of department decides they have a process to automate, so it is done on an ad hoc basis.

Organisations are keen to adopt this new technology, but they don't take the time to review their existing processes; they automate a bad process and then don't get the best results.

A few robots are implemented but not scaled up, and therefore the results of automation are not as rewarding as they could be, limiting the overall impact.

Organisations often completely underestimate the cultural impact on the workforce and how this could be the biggest barrier to a successful implementation. They jump straight in, without taking the time to consult and get everyone on board. Failure to do this can be a major setback.

How has intelligent automation (RPA and AI) impacted you/your clients, and how could it be better?

Benefits:

- *Scale – Growth*
- *Efficiency - Cost savings, Speed, Migrations*
- *Error Reduction – Quality, Client satisfaction*
- *Compliance – Adherence to procedures, policies*
- *Sales – "Forward looking", Resell opportunities*
- *Job Satisfaction – Staff satisfaction, Retention*
- *Others – System/data migrations, Licencing, Staff availability, etc.*

How could it be better?

- *Establish clear owner for the RPA project. Often, conflict of ownership is present between IT & business.*
- *Ensure that IT and the business have good clear lines of communication to enable them to work better together.*
- *Have a definitive overall plan for RPA across the organisation to gain maximum benefit.*

How will RPA and intelligent automation shape the future of work?

The future of work will see us all working in an environment whereRPA and intelligent automation is the normal way of working, following this so-called 4th industrial revolution. Like all revolutions, the role of the people will be paramount, and it will be crucial for company executives to pay attention to the disruption that automation brings to the workforce. Automation will not just be about reducing FTEs and reducing costs, it will be about balancing digital transformation with corporate social responsibility and the welfare of all the employees.

9 Involvement breeds commitment

> "First seek to understand before seeking to be understood",[6]
> Stephen Covey

Empathise

Getting the team managers and end users involved in your RPA journey will keep up momentum and alleviate their paranoias. Furthermore, with the right education, they will inevitably help keep your pipeline full as they proactively identify good opportunities for automation.

---

[6] Book: The 7 Habits of Highly Effective People – Dr Stephen Covey

By having your analysts take the time to immerse themselves into the target team's environment, they can start to understand the difficulties and pain points being experienced on a human level. This way your stakeholders will see that your CoE is there to help them in their daily job, make their work more interesting by removing mundane tedious tasks, improve work-life balance by making them more productive without many late nights, and make their department stand out as they're able to achieve more ambitious targets.

The benefits need to be expressed in ways they can understand and which get them excited and create urgency:

1. Run demos and show case studies of how competitors are using the technology
2. Meet the teams, squash myths and fears, and sell the benefits
3. Understand their drivers and how they are currently incentivised in their roles
4. Understand what they're struggling with, and where their pain points are
    - Where are the issues and inefficiencies?
    - Bring out their frustrations; allow them to relive how bad it is
    - Get them to visualise how it could be much better

Educate

In a company that values lean thinking, even before automation is considered, it's useful to have one or two champions in each team who have taken the White or Yellow belt Lean Six Sigma training course. This is so they understand the importance of quality data and accurate measurements, as well as how data can be used to improve their team's performance and enable them to make better decisions.

To create a culture where staff embrace change to allow your business analysts to work efficiently, the Centre of Excellence team is responsible for providing this basic training on how to identify wastes (we talk about the eight wastes in the next chapter), how to identify automation opportunities, as well as how to collect relevant data needed to assess each business process. Effectively, they can get involved in starting to build the automation catalogue for their team, which can then be reviewed, assessed, and compiled by the business analyst.

If everyone who is feeding the automation pipeline is educated in the same way (and using the same tools), you can be confident that when prioritising opportunities across multiple teams you will be comparing like for like.

Here are some introductory Lunch 'n Learn topics your CoE can run for interested staff:

- An introduction to RPA and intelligent automation (invite a vendor to run demos and show relevant case studies)

- Intelligent automation/RPA Q&A – open the floor up for any queries or concerns

- Lean thinking courses:

    o   Identifying 8 wastes (D.O.W.N.T.I.M.E.)

    o   Organising your physical and digital workspace with 5 Ss (Sort, Set in order, Shine, Standardize, and Sustain)

    o   Tools and techniques to identify opportunities (process metrics, mapping, fishbone chart, and root cause analysis)

    o   The 5Rs of process re-engineering

- Lean Six Sigma White and Yellow belt training

We look at several of these topics and techniques in later chapters

Once your CoE team gets to a level of maturity where you've defined your approach and developed stable processes and have useful material to share, it's worth setting up an intranet for you to start engaging with the organization and for staff to learn more, review previous lunch 'n learn slides, and make requests for your services or ask questions.

10 Valuable assets

To reiterate, it's a very good idea to bring in experts early to set up shop and train your staff. Without a doubt, this will help you accelerate your transformation strategy beyond your competitors. However, it also makes sense to build up your intelligent automation capability from the inside, as much as possible. Granted it's not realistic to train up a staff member to be on par with an AI engineer PhD, and it makes more sense to take on board an experienced solutions architect full time. However, it is does make sense to hire temporary experienced experts who can train your staff to become RPA business analysts, project/programme managers, and developers.

Value is in those who are experienced in both the automation industry and your company's marketplace, be that finance, insurance, or utilities. It's hard to find people who understand how your business works and the dynamics of your stakeholders, so you need experienced staff to find the best way to tailor this methodology to work for you.

The emotional impact of digital transformation

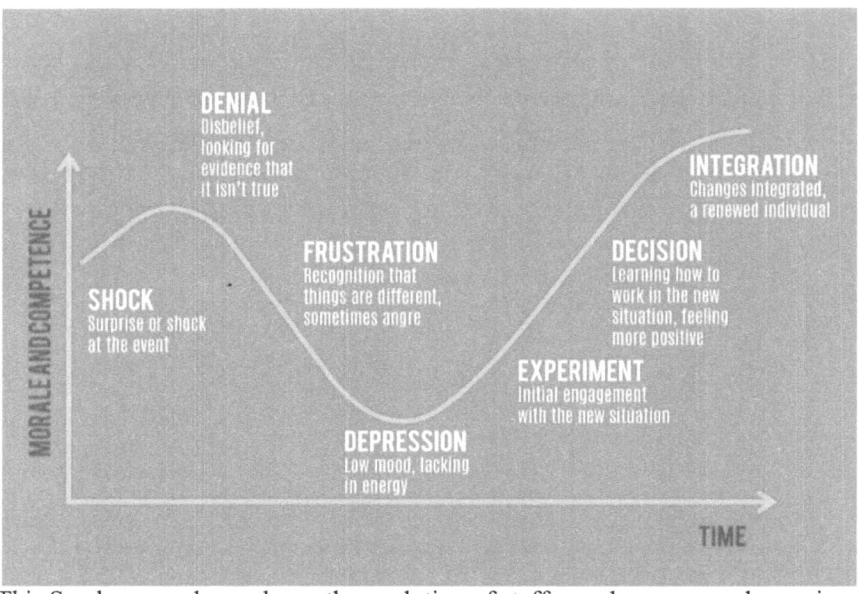

This Sarah curve above shows the evolution of staff morale as a new change is introduced to an organisation[7]

As business leaders, an important stat you need to be aware of is that when companies introduce RPA, 65% of staff feel their roles are in danger. Like a cancer, this can be harmful to your organization as fears spread to two-thirds of your staff. If your company is serious about its digital transformation, it's imperative that you engage with HR and address emotional resistance so that your most valuable assets feel valued.

---

[7] http://centeredbusinessconsulting.com/2016/09/short-term-ceos-the-organizations-failure-to-adapt-to-change/

In addition to the fear staff have of losing their jobs, at a senior management level an important point to note is the way in which companies and corporate cultures financially and emotionally reward managers. Some companies reward staff based on head count that they manage, whereas others may be incentivised by how efficiently they spend their budget (budget vs output). This will understandably have a big impact on how accepting managers will be to automate their teams to improve efficiency, as they may put digital transformation as a low priority. HR also will need to evolve this incentive structure in line with the new digital world before these hidden resistors prohibit your organisation's plans.

Are they educated?

In this section we looked at why it's important to bring in experts early so that any automation initiatives are executed correctly, as do-it-yourself RPA can form bad habits which are harder to change than creating good habits from the start. Test whether your Centre of Excellence team has educated your entire staff:

- Can key stakeholders and staff involved in RPA projects clearly explain what RPA is, and what the benefits are at a business, department, and individual level?

- Have you connected with HR to educate them on the emotional impact of digital transformation with regards to new career paths and incentive structures?

- Has everyone in the teams you've targeted for automation attended at least one lunch & learn session or workshop?

- Do teams understand what the eight types of waste are?

- Do team managers understand what metrics must/should be collected to assess a process's performance and suitability for RPA?

- Does your CoE have its own intranet portal with information, shareable material, and a forum for requests and questions?

Now that staff in your organization have the know-how, it's time to arm them with the tools to empower them.

Scout culture

At this stage, staff in your organisation are excited and believe in the potential of automation, and they understand how it can boost their team's performance. You've run some successful proofs of value, and they've seen the results and heard and shared the testimonials of people positively impacted by these initial bots. Departments are keen to have RPA in their team, as they understand how this can enhance their roles and potentially offer new opportunities in the future.

That's the hard part. Now to keep building the momentum, your Centre of Excellence team needs to start sharing tools, templates, and guidance so staff can proactively get involved—remember, *involvement breeds commitment*. This will also give you the confidence that across the organisation or department, intelligent automation is being conducted in the right way.

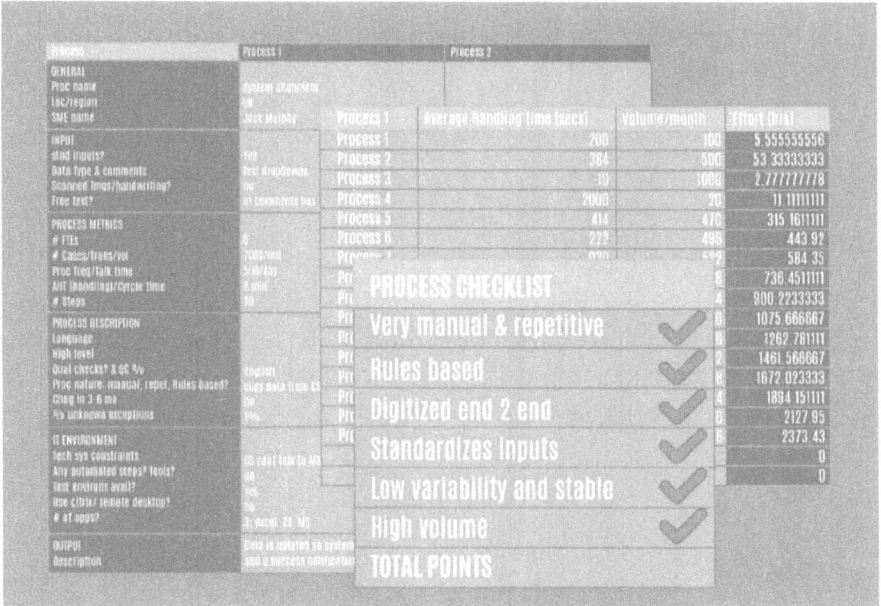

Having 1000 eyes identifying and assessing opportunities is far more efficient than having a handful of business analysts doing all the legwork. Providing teams with tools such as automation catalogue templates and RPA checklists enables departments to start identifying potential automatable processes.

Staff should:

1. List *all* their manual processes (even those that don't seem RPA suitable)

2. Gather process documentation and relevant process information

3. Collect process data, or if available, extract it from a database or excel spreadsheets where this information is stored

The BA is then able to sense-check this with management and SMEs, compile all processes from various teams into a complete automation catalogue, and after further assessment, prioritize this pipeline of work.

I have used this technique of sending templated tools to teams across various departments, working with over 30 senior stakeholders, SMEs, and managers to identify over £2 million of potential savings across an entire business unit.

It's advisable to send a pack which includes some CoE marketing on the benefits to help stakeholders envisage how better work-life could be after the change vs emphasizing how painful and frustrating the current situation really is.

11 Upskill

"Oh, it'll run eventually, and when it does, we have to know how to program it." - Film: Dorothy Vaughan in Hidden Figures (The computers vs the IBM 7090)

Granted, RPA will remove the need for a lot of jobs and reduce the size of teams, as it allows individuals to become more productive. Where a job had required ten people to manage the volume of work, augmented with RPA, it would now only require 7.

So, the big ethical question is, what happens to the other 3?

This seemed to be the most sensitive and generally most avoided question in the media, though of late I've seen it become more talked about as technology advances and automation becomes more ubiquitous. Understandably, every business has a natural attrition rate, due to performance, retirement, or people moving on to other careers. However, businesses should have a plan for how to redeploy their top resource if possible.

- Can they be re-deployed to another team that would otherwise need to hire external staff?

- Will the company itself expand its workforce in the near future, and thus does it make more sense to keep these staff who already know the business, customers, and processes?

- Can we upskill them so they can move into a more technical role to support the new digital workforce?

As your RPA and intelligent automation programme expands, you will need more RPA business analysts, process improvement/lean analysts, developers, or support engineers. There are many business benefits for upskilling those who already know the process, rather than hiring externally. You have a resource pool of people who already have rapport with the staff, they understand the processes and various exceptions that can occur, they've used the applications, and they know the end customers or clients.

As a responsible RPA leader, now is the time to start engaging with HR to review the re-structuring of employee progression paths in line with these new automation capabilities. This not only will mitigate fears of change and job loss spreading through your organization, but will also keep knowledge in-house, and can save on those contractor pennies.

Do they feel empowered?

Verify that your CoE is providing the tools, templates, guidance, and training that the BAU staff need to collectively accelerate your digital transformation strategy, instead of hinder it:

- Has each team interested in RPA been given an automation catalogue template and guidance?

- Is the CoE having regular communication with teams who are pursuing RPA to ensure they are using the right tools, and using them in the right way?

- Has the CoE received any positive feedback or interest from staff keen to learn more about the technology or take on more responsibility to assist in identifying and assessing opportunities?

- Has HR communicated that the CoE may provide new roles in the near future?

In the next charter, we will look at a more hands-on how-to guide from the perspective of the CoE team for identifying, assessing, designing, and implementing your technical solution.

I: Inspect & Ideate

Change needs data to prove it really happened. You can't prove you've realised a benefit if you can't compare before and after. At a minimum, you need to measure average handling times and case volumes flowing through the process to calculate the FTE effort and the maximum potential value (MPV) that can be saved.

Brainstorm what the impacts are and identify where the root cause is (a hint, it's generally upstream from where the pain/problem is experienced). Then, find a solution to fix the problem at its core. Lean thinking ensures your To-Be process is streamlined and thus increases your automation ROI because you're using fewer bots or bot time.

Implementing the proof of concept (POC) gets everyone on side, as they can practically see the benefits and how this technology can positively impact their roles and workloads. Implementation requires detailed accurate documentation of the To-Be process and a clear translation of the business's requirements into a technical solution. Testing is key to ensure a robust solution; however, communication of the new process is also critical so users will be aware of, and happy with, the change

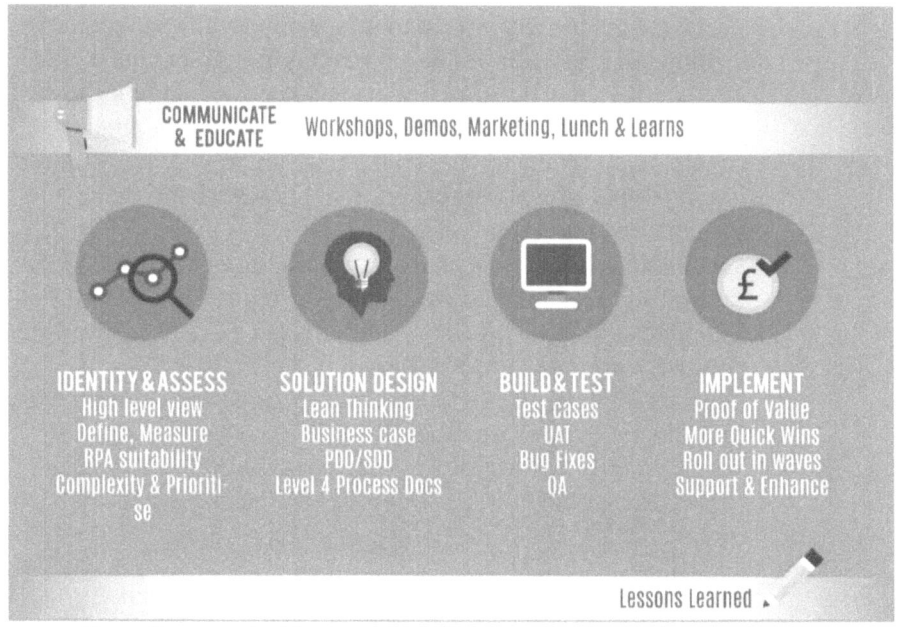

Implementation overview

Here are 10 steps for launching your first pilot. You'll need to hire one or two experienced consultants to set this up as your team shadow and learn on the job so that you can repeat this process in a scalable fashion.

1. Identify the pilot team.
2. Engage with IT and operations management team.
3. Map out the process landscape end-to-end at a high level.
   - Where does the data come from; what do they do with it; where do they send it?
4. Compile an automation catalogue.
   - List all the processes/sub-processes.
   - Assess RPA-suitability (use the checklist).
   - Measure the effort (and potential time saved) = Volume x Average Handling Time (AHT).
5. Build a Complexity Matrix – Complexity vs total benefit vs financial savings.

6. Recommend a POC and Quick Wins.
    - A POC shouldn't take much more than a month to implement. Rolling out a handful of quick wins in your pilot team should take about 2-3 months. Most time is taken unwrapping the team's whole process structure whilst developing a repeatable governance framework as you learn on the job.
    - It's important to make immediate success and build up momentum.
    - Stay clear of large projects that can stall momentum and extinguish excitement, or worse, turn everyone else off completely.
7. Have the business prioritise the pipeline.
    - Use data (complexity vs time saved) to guide and inform their decision and make recommendations.
    - Highlight business benefits for each automation candidate: cost savings, scalability, error reduction, business critical, compliance.
8. Create the business cases for signoff
    - BAs work with SMEs to gather data and process information for the business case: time savings, error rates, volume fluctuations.
9. Create the PDD (process design document).
    - Map out the process to Level 4 (clicks and keystroke level) and label each step.
    - Create a supporting document with description of each step and screenshots to match each step on map (use the same numbering system).
        - A few solutions out there use process mining to map this out automatically (I know—automation to create automation!)
    - Use current work instructions, Skype for business, or Windows built-in 'Step recorder' to speed up this part.
10. Whilst the bot is in development, work with the SMEs (or the quality assurance team) to prepare for testing.
    - Create test cases - Verify the bot works correctly and handles errors.
    - The SME should get the test dataset ready for the bot to process.
    - Arrange for the test environment. Are the applications in the test environment the most up-to-date version?

Zoom out:

"The journey of a thousand miles begins with one step" – Lao Tzu

Wherever you decide to start in your RPA/automation journey, the team or department you decide to target first is crucial for getting the momentum going and increasing your chances of success. The last thing you want to do is start in a team or department which is too difficult or unsuitable to automate, so that you dishearten your team, and your lack of success leads to the loss of faith of key stakeholders that have been sold these shiny new capabilities. A successful launch is even more crucial considering the watchful eyes of those who were initially dubious. They will be watching; all eyes will be on you from the very top downwards, so instant success is your only option in this game.

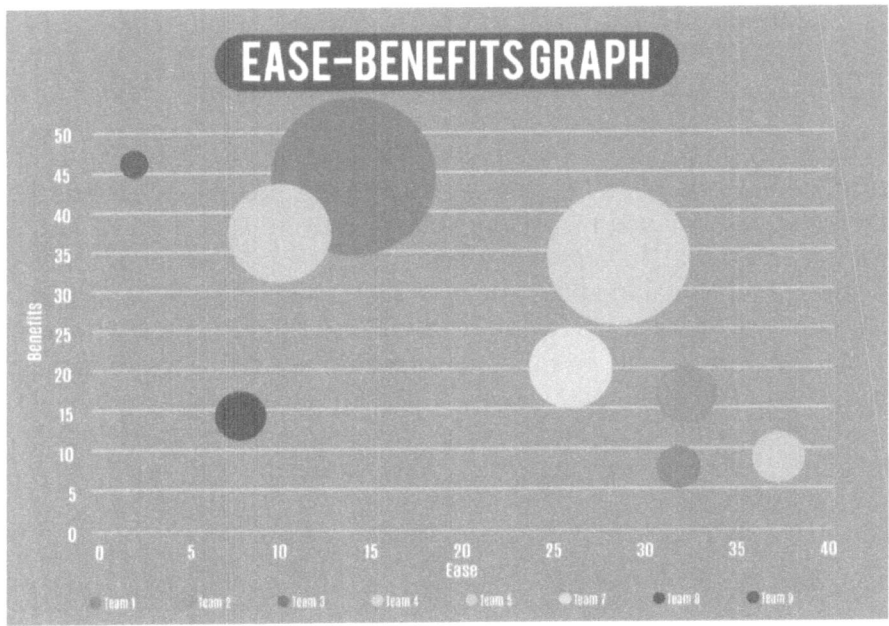

You want to identify the department(s) that are well known to be very suitable for RPA. These are such departments that are full of manually intensive computer-based processes, with large teams of people doing the same tasks. Look for teams with high attrition rates, highly frustrated employees, and large backlogs of back-office work.

You're looking for teams and departments that have a high amount of these types of processes:

- Logical repetitive steps

- High amount of data entry and data validation

- Syncing same data into multiple systems

- Inputting, exporting, migrating data

- Transactional processes handling high volumes

- Errors that have a massive impact on the business

- High percentage of manual errors

- Processes with high queues which delay delivery to customers

- Business areas with high employee turnover

- Large seasonal spikes in work volume

- Processes requiring large teams

- High amount of data searching, data gathering, or data cleansing

- Repetitive updates to databases or form filling

- Moving data from one system to the next, or between multiple systems or databases

- Highly regulated activities

- Financial, compliance, auditing

Avoid teams with these types of processes:

- Low volume

- High variety, multiple ways of completing a task (many exceptions)

- Unstructured data inputs (free text forms rather than dropdown lists)

Types of common departmental processes that you can automate now:

Banking, Finance & Insurance

Appeals processing

Claims processing

Daily P&L preparation

Financial planning

Know Your Customer (KYC)

Loan processing

Logistics – Trade Finance

Reconciliation

Responding to partner queries

Trade execution

Common use cases

Automated receptionist for welcoming visitors to enterprise campuses

Customer onboarding

Daily briefings based on calendar and assigned tasks

Data migration and entry

Extracting data from PDFs, scanned documents, and other formats

Generating mass emails

Issuing refunds

Periodic report preparation and dissemination

Procure-to-pay

Product categorization

Pulling data from multiple websites to identify best deals on auction websites

Quote-to-cash

Transferring business cards to Salesforce

Updating inventory records

Updating vendor records

Customer Service & Sales

Automating multi-step complex tasks that require little decision-making

Creating and delivering invoices

Obtaining detailed billing data

Loading a detailed customer profile

Resolving simple but common customer issues

Updating user preferences and other user information

HR

Absence management

Candidate sourcing

Employee data management

Employment history verification

Expense management

Hiring & onboarding & headcount reduction

HR virtual assistants

Payroll automation

Tech & Support

Fault remediation

Opening up internal tools to customers or employees

Regular diagnostics

Regular testing

Software installations

12 EAR (Enterprise Automation Road Map)

"Friends, business leaders, analysts, lend me your EARs" – adaptation of Mark-Anthony in Caesar.

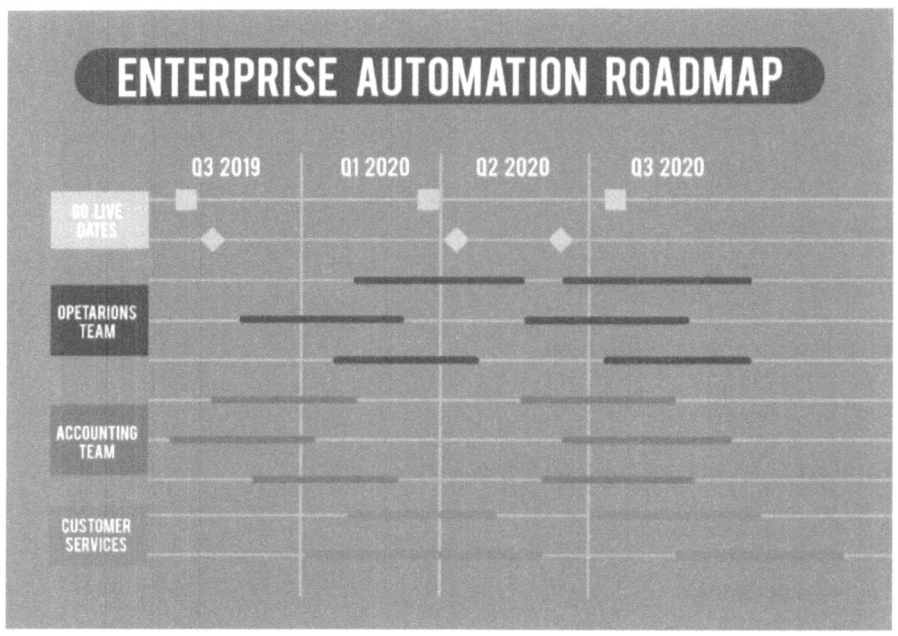

Now then, you've had a few successful pilots, overcame the challenges, and learnt to avoid certain pitfalls. You've even prioritised processes in your target team and have probably zoomed out to prioritise other teams and departments across the organisation at a macro level. Now it's time to map out your digital transformation plan.

The enterprise automation roadmap (EAR) is the holistic plan of how to roll out automation throughout the enterprise at scale. It is a timeline of when you plan to implement each wave or group of processes, and how you plan to target other departments or use new technologies.

Here are five steps to building your map:

1. **Zoom out:**
    a. Identify the departments most suitable for RPA
    b. Solution layer: Review and vet capabilities and vendors that can meet individual team needs (RPA, AI, machine learning, intelligent character recognition, process mining, business process management, etc.)
    c. Build a department complexity map to determine which departments or teams are easier to automate and which have the most value
2. **Zoom in:**
    a. Identify all suitable processes and add to your automation catalogue
    b. Solutions layer: Review and vet capabilities and vendors that can meet individual process needs (RPA, AI, outsourcing, lean process re-engineering, etc.). Which solutions have the ability to:
        i. Create libraries of re-usable code
        ii. Scrape web pages, Citrix applications, PDFs
        iii. Integrate with plug-ins and APIs
    c. Build a process complexity map for your first target department. Identify the Quick Wins
3. **Categorise and prioritise**

a. Prioritise teams and processes into waves using your high- and low- level complexity maps

Using a complexity matrix, you can capture the difficulty of each automation from the developer perspective. By quantifying the complexity and intangible benefits such as scalability and error rates, you can plot all processes on a complexity/benefits bubble map. Adding effort (volume of cases x AHT), which is the potential financial savings, as a 3rd dimension (the bubble size) will give a robust view on how to prioritise.

However, doing all quick wins upfront may not have significant impact on the business, so it is ultimately the business who should decide, guided by the organisation's strategy and using this map as a decision aid which is backed by data.

4. **Bake in continuous improvement**
    a. Keep coming back to implemented automations to improve or enhance them
5. **Plan how to self-fund your CoE**
    a. Use savings generated from implemented processes to pay for the continued running on your CoE—that means get profitable fast, be that ROI from clients or wooden dollars internally

The benefits from previous automation projects should pay for future initiatives. As I hinted at previously, each wave of automation development should be a mixture of large and small RPA projects, not too many small projects which deliver minimal benefit, whilst ensuring that you do not focus too heavily on the big projects which would stall momentum and kill excitement amongst stakeholders.

Delivering large benefits from RPA will inevitably pave the way for your organisation being willing to adopt AI and machine learning capabilities, thus opening up a whole new world of automation opportunities.

Identify and Assess

"If I had an hour to solve a problem, I'd spend 55 minutes thinking about the problem and 5 minutes thinking about solutions." – Albert Einstein

You can either automate the existing process, or you can optimise it first; however, I believe in GIGO (garbage in, garbage out). It's advisable to really understand the problem and look at the details. This will help you prioritize which problem to solve first, and, if you measured before you implemented a solution, when you measure it after, you can confirm that the benefit has been realized.

If this is your first attempt at RPA, first identify a pilot team who have highly RPA-suitable processes and are most keen to be the guinea pig.

Identify the problems and pain points (high-level view)

- Start by meeting the senior manager or head of the department you will be automating. Understand their general pain points from a high-level view.

- Set the expectation that this will require SME time for interviews and workshops to map and understand the process and its metrics, as well as design solutions. Buy-in from the top will result in the removal of barriers and give management advanced notice to free up their team's time.

- It may help to map out the high-level process architecture end-to-end to see where the pains are *experienced*, and identify what the up-stream processes are (which may be where the actual problem lies) and see if there are any other downstream impacts as a result.

Identifying the right process is the be all and end all. Automating processes which are too large or too complex at the start will kill the momentum your team have worked so hard creating. However, if your first few processes don't produce enough benefit or solve key issues, you'll quickly lose the credibility of RPA.

13 Define and Measure

- List all manual processes (and sub-processes) that the team perform.
    - This might be a long list, so IT may be able to assist in extracting this information from the system applications. However, as many process users do 'swivel-chair' tasks (which involves moving data between multiple applications), it may be difficult to do this.
    - You can use process mining software that sits on staff computers and watches the actions of users to automatically create a process map and records actions to collect metrics.
    - If you've managed to create a Scout culture, then team members can assist with collecting this information.
- At a minimum, you want ensure that the monthly volumes and average handling times (how long it takes to complete one iteration through that process) are collected.

To have a logical approach to choosing processes, you will need data to back this up, which is another reason to involve the SMEs to gather or sense check findings.

14 RPA suitability

IDENTIFY, ASSESS, PRIORITISE

o Put processes through an RPA suitability matrix, which is a simple checklist of questions to see how suitable the process is for automation.

    o For each process rate out of 5 (high-volume, rules-based, etc.), some points are not merely a Yes or No but are on a scale, and so your final score will indicate the complexity and benefit potential of that process.

    o Filter out processes that don't meet the minimum requirement for RPA.

    o Filter out those that don't meet your threshold (your threshold will depend on the type of process mix you have, the maturity of your CoE, and your objectives). Perhaps you want quick wins to focus on savings and getting buy-in. Or developers are inexperienced and you want to get started quickly. Or you're a mature CoE and your objective is to capture all processes, including the more complex ones.

- Put processes in your automation catalogue: this is your complete list of all the processes you aim to automate. This isn't to say that you won't circle back after you've completed the list and gained more capabilities (we'll cover this later), but this gives your team focus and gives leadership indication on what will realistically be saved after the automation catalogue has been completed.

15 Focus with the 80:20 rule

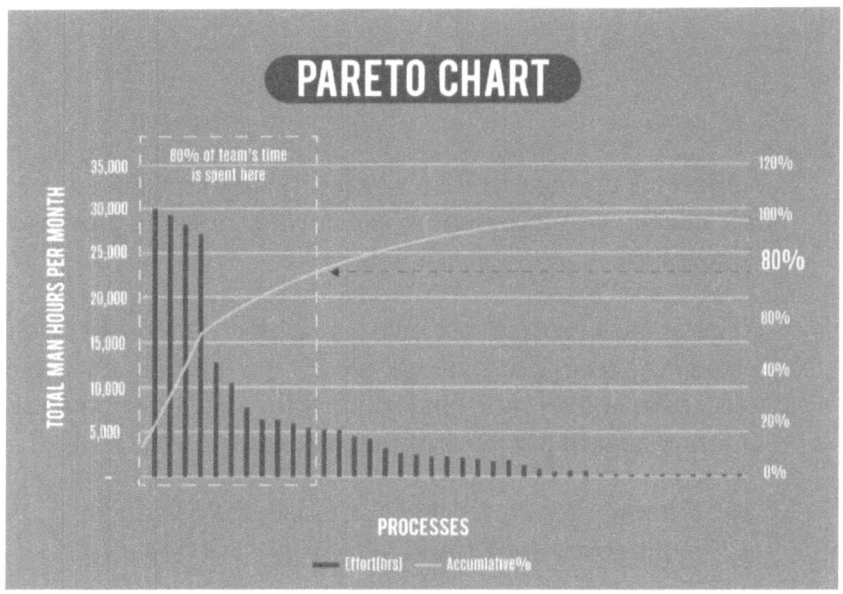

When your automation catalogue has hundreds of processes, it will be difficult to visualise all of them in a presentation, and it can be overwhelming to decide where to start, so it's advisable to create a shortlist using some logic, and then you can dive deeper.

- Put your automation catalogue of processes into a pareto chart. Sort by the effort of each process.

- This shows where 80% of your workforce effort is, which is generally in only 20% of all the manual processes the team performs.

16 Process complexity

With a shortlist, you can analyse these processes closer and collect information that will help you understand the complexity and ease of implementation.

This complexity assessment will assist your developers in understanding the level of effort required to build the bot, and hopefully if your Center of Excellence sub-teams are aligned, your support team will also be able to use this to understand at a glance the effort needed to maintain the bot.

To minimize the disruption your business analysts cause the BAU teams, document analysis is definitely the first point of call. If the team has work instructions (which include screenshots), this can help answer many of the questions that your analysts will need answered to understand complexity.

- Is input data structured? Are processes standardized?
- What types of unstructured data are fed into this process (scanned images, handwritten, or free text)?
- How many FTE hours does this process take each month?
- What are the average monthly volumes, and are there any cyclical variations or forecasted increases in volume?

- How frequent does this process run and what is the Takt time?

    Takt time = start of one unit entering a process to the start of the next unit

- What is the average handling time?

    AHT = time for one unit to completely run through entire process (including waiting time)

- How many screens and applications does the process use?

- How many decision points and steps does the process have?

- Are there any control checks or 'man-in-the-loop' checks?

To ensure you cover all bases, use the **SIPOC** framework to think of all the aspects of a process:

S(Supplier): Where did the data for this process come from and how was it received?

I(Input): What does the data look like?

P(Process): What steps should the bot have taken?

O(Output): What does the transformed data look like?

C(Customer): Where did the data go and how was it sent?

We'll go into SIPOC in more detail in the Lean thinking and Support chapters.

For the more detailed process metrics questions, just looking at the documents and speaking to SMEs might not be a true reflection, so you're going to need to zoom in. If you're short on analyst resources, you may want to get a rough estimate so that you can focus further on a handful of bigger ticket items.

17 Prioritise

Many RPA teams order their processes by the amount of effort savings (volume * AHT) or financial benefits but stop at this point. It's advisable to go two steps further and order by all benefits (including intangible benefits) and ease of implementation, which is a three-dimensional prioritization.

Plot processes on a bubble chart by:

- **Effort** aka FTE savings: Volume x average handling time (the size of the bubble)

- **Benefits**: Monetary benefits and quantifying intangible benefits that may indirectly impact finances (removing errors or ensuring compliance which would mean avoiding fines or losses)

- **Ease of implementation:** Using the score from your complexity matrix, you can determine how complex a process is (this would require the business analyst and SME collecting more data)

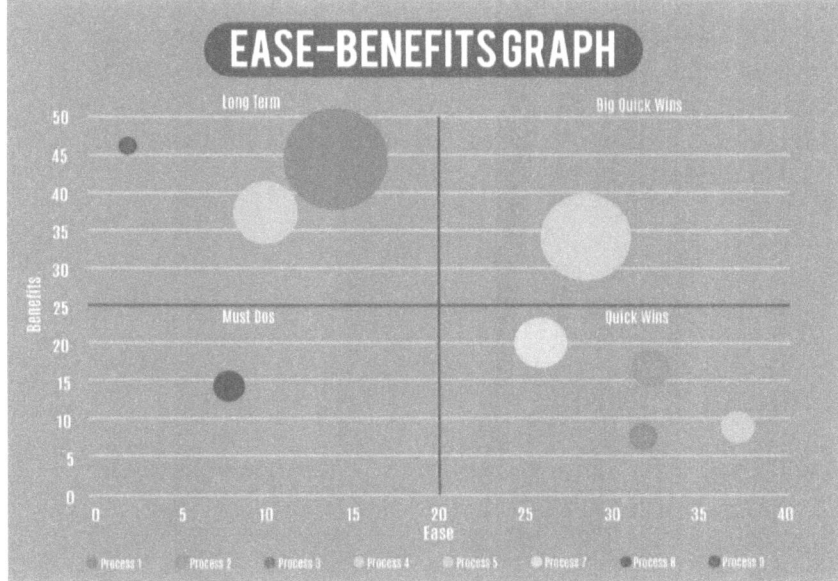

The above chart shows four quadrants which demonstrate to leadership where the easy processes which provide big benefits are, the quick wins, the large benefit but complex processes, and those that are difficult with minimal benefit.

Potentially, you can work on the groups of processes in each quadrant one at a time, starting with quick wins, then big tickets, etc., though this is not advisable. Besides, it's up to management to decide on priority based on their strategic objectives (perhaps best to start with a mixture of very quick wins and larger processes that give quick and substantial benefits). Group the rest of the processes into waves or phases.

As you can see, as your CoE matures, this map can also expand to show other automation and AI solutions on the map together with RPA so you can pick and choose to match your budget and objectives.

Intelligent automation catalogue

Senior leadership team (chief operating officer or operations director) can pick which solutions they want for addressing each root cause or pain point. Using a costing model, they can select and change various solutions options to calculate different ROI outcomes (e.g. if they have a £500k budget to spend on the CoE team, the analysts can model different scenarios).

18 Zoom in further:

SMEs will have a gut feel for many of the questions on process complexity to a fair degree of accuracy. However, you'll get a great deal more information from documentation analysis first. So, use both document information and SME feedback to ensure there aren't any gaps when building the process map.

Document analysis

As discussed in the challenges of RPA, what is documented (and sometimes you'll be lucky if process documents even exist) doesn't usually reflect what is reality. It's generally an old version that hasn't been updated for months or years. Even if the document has been kept up to date, it's worth noting that the people who create these documents know the process so well that they can leave big assumptive leaps between steps, which is a cognitive bias called the *curse of knowledge*. Having your business analyst, who doesn't know the process, go through this is beneficial both to your RPA team, as you will get a clearer understanding of its automation potential, and the team receiving the revised document, as the document is more fit for purpose when it comes to training someone new (be that training on interacting with a new automated process, or the current process if it doesn't end up being automated).

Start by reviewing work instructions documents or how-to guides that the team use to train staff. This will help you understand the steps along the 'happy path' (the fastest, most preferred route through the process). However, some documents may also show where the decision points are and may provide steps for handling exceptions down these alternate paths.

Document analysis is a preferred place to start, as it's the least disruptive method of gathering information.

Interviews

Interviews with SMEs are very disruptive, as they require taking SME(s) away from their day job to answer questions. It's best to ensure that your analysts are prepared from analysing documents prior to the interview, and the interviewees are prepped with the topic of discussion and have had time to collect any information required to bring to the interview. This will keep up the speed of the interview and reduce chances of needing to re-interview the SME again. As every business analyst knows, it's best to stay away from closed questions, this way the SME will elaborate on the context of the process as well as expressing any concerns they have.

Shadowing

As previously stated, documentation and even SMEs inputs may not necessarily reflect reality, as users find shortcuts and workarounds, so to go one step further to see what they really do, have your analyst sit with the SME as if they were their apprentice and allow them to train your team. Let them first walk and talk your analysts through the process and the applications so they understand what they are seeing (as SMEs are pros at this, they may need to go a lot slower than their normal pace).

Asking 'why' questions encourages SMEs to talk out their thought-process as to why they made certain decisions. Your analyst can sit and map this as they watch and annotate whether decisions are logical/rules-based or subjective. They can also start to map how exceptions down different paths on the decision tree are handled.

If possible, it's advisable to screen record the process and the SME's voice as they walk the analyst through the process. This way the analyst can refer back to the recording to clarify anything missed, again reducing the amount of disruption to the team.

Realise that different SMEs may do the process differently, so sitting with a few people (especially if it's a big team) will give you great insights, but remember to find out the best way to handle the process and then make that the standard.

Integrated workflows aka swim lanes

To understand how information is passed from person to person or system to system, an integrated workflow diagram is the best way to depict this. Compiling all that you've learnt about the process into this diagram will be a powerfully simple way of seeing where the wastes are (see Lean Thinking chapter). This can also help you ensure that no steps are missing, as it will be apparent if there are no logical segues between adjacent steps.

Once mapped, and revised, process metrics such as volumes, AHTs, straight through processes and automation percentages can be highlighted on the map to give further meaning to what is being shown.

Again, some process mining and process discovery software can assist you in showing this

Being a 'Business Analyst' in a small team

Working as a process automation business analyst is very rewarding, especially as I've building my career in growing businesses who see benefits other than FTE saving, such as reducing errors, increasing speed of service, reducing queues, enhancing user experience, and creating scalability, to name a few.

Even when teams didn't require as much FTE, staff (who are the real business value, with their immense knowledge of the business process and industry) had the opportunity to be upskilled or could be re-deployed to other faster growing business areas.

The worst mistake a business could do is remove staff, only to rehire new staff months later.

I moved into RPA because I saw the potential of process automation and how powerful automation and Lean Six Sigma could be.

As a Lean RPA analyst/project manager, I'm not here to bring the big ideas; the ideas and solutions come from the SMEs and end users, because no one knows the processes better than they do. They need to be deeply involved in designing the change, instead of change being imposed on them. I truly believe that involvement breeds commitment.

I'm just here to understand their pain points, define their requirements, and share my knowledge and passion of process automation and optimisation (Lean Six Sigma), to give recommendations, guidance and support to help teams achieve their ambitious business objectives. From developing tools, re-engineering processes or re-designing strategy

I've worked in Aerospace, Energy, Insurance and Fintech.

My goal is to help business leaders improve their team's productivity by ensuring two things: optimised processes and a motivated team.

Productivity = Optimised processes + A motivated team

So, using process automation and Lean Six Sigma, I work closely with teams to uncover ways in which we can fix their workflows by removing technical road blocks and mundane time-consuming tasks that frustrate them and get in the way of their daily activities. Some of these time-consuming tasks include data migration, filling forms, 'copy & pasting' data between systems, and other things that need to be done but get in the way of their actual job, where they add real value.

If we can automate these tasks, teams can be more productive and achieve targets and deadlines faster so they can focus on the more interesting, creative, and value-adding work. Teams are happy because they have less late nights so better work-life balance, and heightened job satisfaction. The business is happy because it can grow faster and give an enhanced customer experience.

What does an RPA business analyst do?

As the BA, you're the liaison between the business and the development team. The BA uses tools and techniques to elicit requirements from the business, and has good understanding of the technology (preferably enough to build or edit bots) to advise the team on what's possible, and what can't be done.

The business analyst is responsible for development of the PDD (process design document). This is the most important document because it translated the business problem into a high-level technical solution. The BA is the intermediary between the SME and development to ensure that the document has everything the developers need to build the required bot.

Once it is built, you need to ensure that the UAT is successful and to validate that it meets all requirements. This means that, ahead of time, you prepare test data and test cases with the SMEs.

So it is crucial that you ask lots of questions and delve into the detail. This may uncover hidden requirements.

Since the RPA analyst plays such a central role in the CoE team, if the CoE is small or new, a versatile RPA analyst may need to wear several hats. Here are 10 things they may be expected to do:

1. Assist in developing analysis best practices for the Centre of Excellence team.

    a. Help develop and ensure the correct use of tools and templates for identifying and assessing opportunities.

2. Be an evangelist for RPA—educate and upskill teams on RPA and lean process improvement.

    a. This could include running lean and RPA 'Lunch & Learns' to ensure that teams are not just bought in but are involved in identifying, assessing, and re-designing processes.

3. Identify and assess automation opportunities.

4. Gather and analyse requirements, and write up use cases and test plans

5. Facilitate root cause and solution design workshops with stakeholders and process experts.

6. Have enough knowledge of the RPA platform(s) to, at the very least, be able to build a basic automated process. This is so you are accurate in advising the business on what is and isn't possible.

7. Help develop an implementation roadmap for RPA, as well as suggesting other automation tools (scripts, macros, commercial off-shelf tools, etc.)—this will inevitably lead to working with other technology like OCR/ICR, machine learning, and AI.

8. Work with subject matter experts to document As-Is process maps and procedure, and run process re-design exercises for To-Be solution design.

9. Record users executing a process and develop keystroke documents for the process design document (PDD)

10. Support UAT and evaluation of automated processes—make bug logs for developers to address.

11. Provide continuous support for automation improvements and enhancements.

One final point: In many cases, you may also need to take on the project manager role to ensure that you and your team achieve objectives and meet deadlines.

Inspect: Have you identified **and** prioritised opportunities?

Presenting the makeup of all manual processes in a Pareto chart and/or complexity map is a clear way to depict what the automation opportunities look like across a team or department, and it guides decisions.

Review these questions below to confirm that your team has prioritized your automation candidates in a logical way:

- Have you measured your processes by effort (AHT x volume) and potential saving?

- Do you know which handful of processes make up 80% of all your team's effort?

- Have you reviewed your top processes in detail to understand their complexity?

- Have you plotted your top processes by Effort, Benefit, and Ease of Implementation on the Complexity quadrant?

- Have you provided the leadership team with a visualization of the makeup the department's processes so that they can strategically prioritise which processes to automate first?

In the next chapter, you will learn…

What to do with the opportunities you've identified. You will decide if you should automate first or apply lean thinking. You will learn how to find the right solution and get a higher ROI.

Solution design is like herding cats

Two schools of thought:

If you've been in the RPA world long enough or been to a few conferences, you'll notice that there are two main approaches businesses are taking to RPA.

Lean then automate …or just automate (aka GIGO)

Many organizations, and some vendors and consultancy firms, focus on automating existing processes as they are, taking exactly what a user does and automating it. Understandably, the method is a fast fix to show results and deliver short-term value, and the obvious benefit to the business is an immediate savings with little fuss about how the process could be better. No time is wasted on workshops or process re-engineering; this approach is not too disruptive, and just provides like-for-like automation.

The problem with this is that there can be inherent risks and problems with the current process design. A bot can make the same mistake as the human but would do it 1000 times in a split second, or it can cause a bigger strain on bottlenecks due to the increase in throughput. Also, the ROI would be suboptimal, as solving multiple downstream problems could have been better solved by focusing efforts upstream where multiple problem originated.

It's apparent by the name of my consulting firm, Lean IA (Lean Intelligent Automation), that we favor the latter. Naturally, as a Lean Sigma black belt, I like to understand root causes, identify wastes, and facilitate the design of solutions to improve the efficiencies of the underlying process. To me, as to any lean process improvement consultant, RPA and AI and other automation tools like BPM software or automated workflows help remove wastes and are powerful enablers to optimise processes in order to deliver more value for the business and to the customer.

To this end, the following are techniques to dig deep into the problems and take stakeholders along the solution discovery journey.

Workshops

Workshops are probably the most powerful way to extract information and get great minds together to create a viable solution to solve current problems your business is facing. The other reason that it's the most popular method is because you get a cross-pollination of ideas, and if managed in the right way, you get 100% participation, without having one person (normally the most senior) overpowering the room. The biggest challenge is getting a group of busy people in a room at the same time, so it is good to mitigate this by giving a lot of advanced notice. Furthermore, it's helpful if the RPA sponsor prompts department heads to encourage their staff to attend and view these workshops as high priority.

It's recommended to split this into two workshops, a Root Cause workshop to identify where the real underlying problem sits and what is really causing it, and then a Solution Design workshop to create a way or ways to minimise or fully remove this issue.

19 Root cause workshop

It's imperative that you involve all key stakeholder groups who are directly or indirectly affected by, or potentially causing, this problem. So, understand where the problem/pain is being felt, and then look at upstream and downstream teams to find the right people to invite.

The objective of this workshop is to first brainstorm a list of problems being experienced in your target team, and then using root cause analysis (RCA) techniques, create a handful of root causes which are responsible for all those problems. You also want to use this time to validate data you've found in documentation or been given by a SME and to clarify the end-to-end process.

By validating this in a group, hopefully someone other than the original source can sense check this, but what you may also discover as a group is that the different teams (upstream, target, downstream teams) may have several misconceptions of what each other do and how they do it. The reason for this is that, generally, companies work in silos, and outside the process of transferring information, they don't usually communicate in a way to better understand or update their understanding of how things are currently done. This leads to teams blaming and not re-aligning with each other, and as a result many problems are created. Just having the right people in the room talking about what they do in their part of the end-to-end process can quickly uncover solutions. However, ensure that solutions are documented for later but are not discussed at this time, as the focus is to find the root cause, so prohibit solutionising at the stage. This is simply because you do not only want to solve the problem, you want to solve it at it's root.

I'm assuming you've invited exactly the right people that you need, and that they are not just present because they were available (don't fall into this trap), but because they can add real value into the conversation. If that is the case, then you need to ensure that everyone's voice and opinion is heard, which requires strong facilitation skills and techniques such as 'round the table', so everyone gets a turn to speak, or 'silent post-its', so everyone writes their ideas/issues/opinions on post-it notes for the whole group to discuss.

When guiding the group to uncover potential causes to the problem being experienced, Lean Six Sigma uses what's called an Ishikawa (or fishbone) diagram. Some may also refer to this as the cause and effect diagram, as the six fishbones are the six causes of sub-par quality, and the head is the effect (the problem).

The six causes are

Measurement: Are you measuring performance and quality poorly or not at all?

Materials: Does the team have the right resources to do their job properly, and is the working space acceptable (materials are resources, not to be confused with machines, which are tools)?

Methods: This is the process itself. Is there anything inherently wrong with the process?

Mother Nature: The environment—physical, economic, social, legal environments.

Manpower: Your workforce, people involved in the process.

Machines: Tools and machinery used for the process, like the computer or applications.

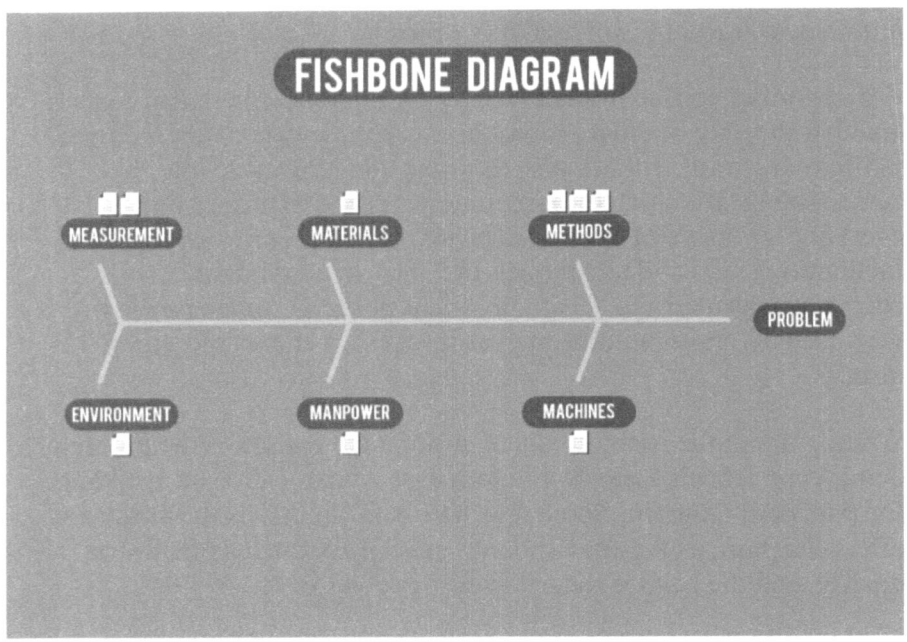

20 Solution design workshop

In this workshop, it may be useful to include all the attendees of the root cause workshop for consistency, but now it would also be advisable to include technical advisers to help the group understand the art of the possible. That may be experts in Lean, RPA, and IT; however, there are many other solutions which may be a better or more logical fit than RPA.

Here are some solutions to consider:

- Lean (always consider this first)
- RPA
- AI
- Application development
- Scripts
- APIs
- Macros

You want to have a few solutions for each root cause. Then you can sort these by what is in or out of scope; you can take this away and calculate development costs, time to deliver solutions, and ease of implementation. These can be fed into your complexity bubble chart, which gives the leadership team a much clearer view of their options (sometimes RPA isn't the answer).

Ideate: Have your solutions involved all the right people?

When creating the solution, hopefully you see the importance of including those stakeholders that may cause or be affected by the existing problem and may be impacted by the solution when implemented. Take yourself through the following questions to ensure you've covered all bases:

- Have you involved key stakeholders from the team(s) that are upstream from your process, who feed information into your target team?

- Have you involved key stakeholders from the team(s) that are downstream from the process, who receive information from your team?

- Do all the root causes identified by the workshop group relate to all the problems being experienced in the target team?

- Have you first considered lean process re-engineering?

- Did you have a technical expert present in your solution design workshop?

- Did you have full participation with everyone's opinions being heard and respected?

Stakeholder management

One more thing, before you set this automation beast loose: ensure you have a strong change manager in place. They are there to keep the gears well-oiled to ensure your RPA/Automation factory keeps delivering solutions at a steady rate. They ensure that changes to the BAU environment are controlled, the production environment is protected, and the staff and stakeholders involved remain onside.

Remember, about 65% of your staff will feel their roles are in danger. Are staff feeling threatened? Are they feeling audited by your analysts? Are they unsure on what this tech is, what the benefits are, and why these changes are happening? If so, go back to "Educate & Empower" and strengthen this up.

Again, have you and your RPA leaders engaged with HR to review the re-structuring of employee progression paths in line with these new automation capabilities? Showing your best staff how they will progress alongside automation, RPA, or AI will reassure them that their livelihood is not in danger and encourage them to stay or become even more productive as new exciting opportunities lay ahead for them.

In the next chapter…

We're going to circle back and take another look at solution design and how to optimise your processes. We mentioned in this chapter about the two schools of thought regarding automation (automate first vs lean process first), and now we'll take a deeper dive into lean thinking.

O: Optimise

Your team have identified the opportunities, prioritised them, and obtained funding from the senior leadership team to go ahead with your designed solution. It's time to implement your solution(s) and optimise your business processes, but remember to keep thinking GIGO (garbage in, garbage out). Continuous improvement must be baked into the plan; automation isn't a one-and-done fix, it's a continuous revision of the process.

Optimisation goes together with Ideation, ensuring that the To-Be process removes as much waste as possible—that is, the eight wastes of Lean Six Sigma: Defects, Over production, Waiting, Non utilized experts, Transportation, Inventory, Movement of people, Excess processing, and also removing as many non-value-added activities as possible. Let's dive deeper into what Lean Six Sigma techniques we can use.

21 Lean Thinking in Solution Design

Lean = Providing high value in the most productive way possible (speed), six sigma = providing high quality as consistently as possible (accuracy)

Lean Six Sigma is a powerful skill which can significantly enhance the benefits you get from automation. It's about understanding how to deliver consistent customer quality in the fastest possible way. Lean will help you increase value from your business processes by removing non-value-added activities and reducing variation and overloading of resource.

By optimising your processes and positioning your automation 'bots' in the right points, you can make a lot more savings with a lot less bots. In this chapter, we'll look at some techniques to identify wastes and sub-optimal parts of the processes to find and solve root causes in the most efficient way. This is an overview of the techniques your Centre of Excellence team need to become familiar with. However, I strongly recommend that they take further education in Lean Six Sigma and Lean Thinking courses.

D.O.W.N.T.I.M.E.

Eight forms of waste exist in your business and in your team right now. No matter what type of industry or business you are in, these forms of waste likely exist in every department.

DEFECTS: Missing information, missed deadlines, incorrect versions

OVERPRODUCTION: Too early, too many, non-priority items

WAITING: Delay between steps and processes (queues, unread emails, piles of paper)

NON/UNDER-UTILISED EXPERTS: SMEs not involved in improvement activities

TRANSPORTATION: Unnecessary movement of documents, data, materials

INVENTORY: Providing more than customer needs

MOTION: Needless movement of people (keystrokes, screens, applications, paperwork, chasing paperwork)

EXCESS PROCESSING: Doing more than customer wants (too many reviews, red tape)

RPA and AI are tools to help reduce or remove these types of waste. However, process re-engineering is what will help you solve this in the least wasteful way possible before implementing such tools.

For example, you may have a team that scans paper forms received from multiple clients and saves these images into a database for their records and auditing purposes, and then the team manually types in the details from the form into your CRM database. A kneejerk reaction may be to get ICR software, which uses OCR software to turn the scanned image into text, then uses NLP to validate and interpret information (perhaps the OCR read *LDONON* as the city, so the AI software understood this to be *LONDON*), and then you use RPA to take this now structured data and input it into your CRM database.

This sounds great, but with lean thinking as a first point of call, it may actually make a lot more sense and be more cost-effective to take a step back and create a user interface for clients to input their details in a form that is added directly to the database, rather than sticking automation over a bad process like a band aid.

SIPOC

Now that your team understands what inefficiencies look like, it's time to find where these wastes are hidden.

The SIPOC model is from Lean Six Sigma and is a great way to understand the flow of information through a process at a high level. In addition, in RPA it helps to understand how data flows through your systems, from team to team and person to person. **S**, for Supplier, is where the information came from, **I**, Input, is what the data looks like and how it is received into the targeted process. **P** is the process you are analysing—take note of how data flows step by step to transform the input data into something else. **O** is for Output. What does the new data format look like and how is it sent on? **C** is the Customer, who is the end recipient of the data/information that the process produces.

[8]Here's an example of that SIPOC model. The bottom half is where process metrics and other information are collected. This is where we look at the volume of cases passing through the process, average time it takes for one case or unit to flow through the entire process, how many FTE are available, what the input and output error rates are, and so on.

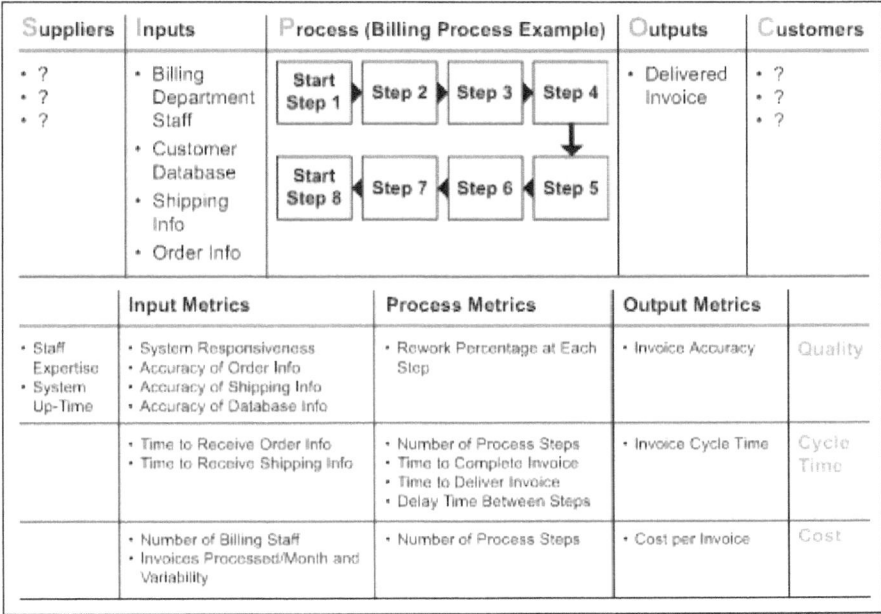

https://www.pinterest.co.uk/pin/203858320612682072/

VSMs

Value Stream Map (VSM) is another way to represent a process to expose where the value-added and non-value-added tasks are. These up/down lines at the bottom show the downtime between each task, this diagram also looks at inventory, available FTE at each point, cycle time (how long each step takes to complete), and the various communication routes between each component of the end-to-end process. Note that the VSM only maps the happy path at a high level so that it's a linear flow, unlike an integrated workflow which can map various paths.

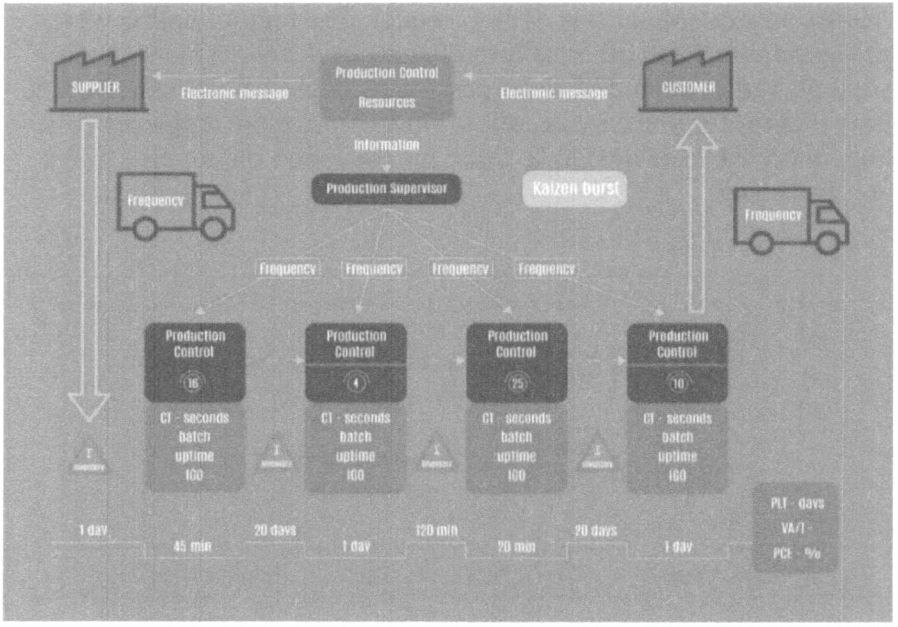

The two biggest wastes that are most obvious here are Waiting, and another waste tied closely to it, Inventory. Inventory is when a business creates more than the customer demands so you have a supply surplus. This could be in the form of piles of paper, long calls in queues, or even tasks queued that are waiting to be actions by a robot. Generally these two wastes go hand in hand. If there is a delay in your process due to a bottleneck or a poor balance of resources, components (be that applications, staff, or your digital workforce) will need to wait on others to pass their data on, and this will result in a growing backlog, or inventory that cannot be processed as soon as it is ready. Ensuring you have good balance of resources across your entire process will ensure that you keep your information flowing and your inventory to a minimum, enabling your business or department to run like clockwork. Now we'll look at how to do this.

With the process clearly displayed, you can create a future state map to visualise how the process could be improved.

Takt time

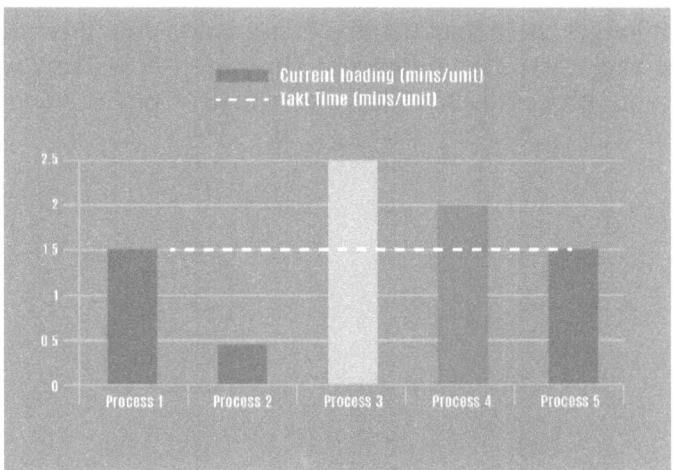

This is an interesting chart, and it may need a second look to get your head around this, but it will help you understand how fast you need to be processing units (widgets on a production line, orders or customer queries, or even cases through an automated system) to keep up with customer demand, and how to meet this demand.

The Takt time chart (Takt literally means the "pace" or "rhythm" of your process) involves three measures: cycle time, lead time, and total time available.

Cycle time is how long it takes to complete an individual task of a process.

Lead time is how long it takes to complete the entire process end-to-end (i.e. how long it takes one unit to past through the whole value stream e.g. call to cash, from call to getting paid).

Total available time is how much time your resources has available in total (e.g. four people working 8 hours a day have 32 hours total, whereas 1 robot may have 24 hours of available time per day).

Using this chart, you can understand how an overloaded component of your process can slow down your whole operations. If one task takes a lot longer to do than the preceding tasks, then this will create a growing backlog, i.e. a bottleneck. This chart shows how well-balanced your process is and identifies which step(s) are over or under loaded.

Takt time = total time available / total volume (or demand) in that same period *e.g. four workers are available 8 hrs a day and one bot is available 24 hours a day. They receive 2,240 cases to process, TT = 3,360 mins / 2,240 = 1.5 mins/case*

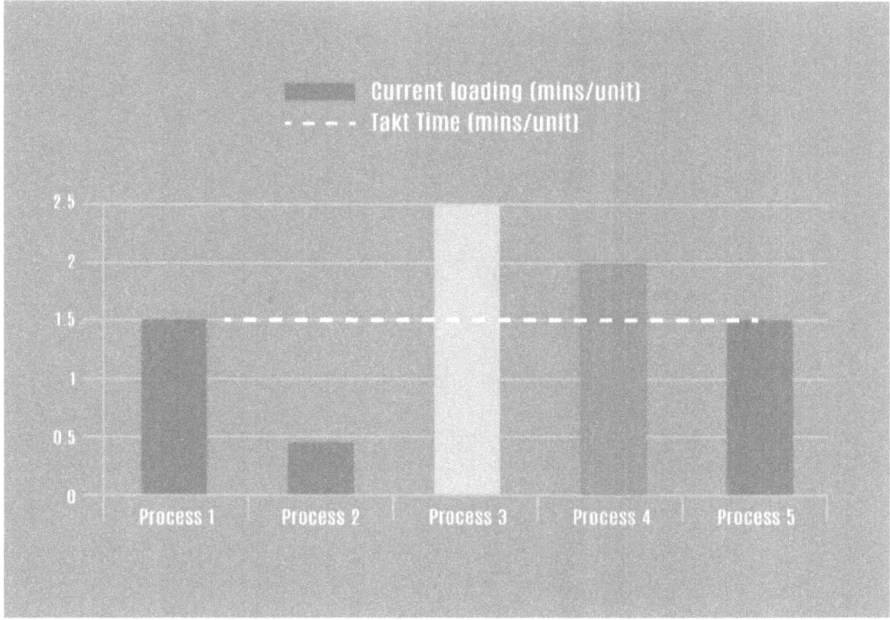

As you can see in the diagram, a few tasks are above the Takt time line and some are below. Person 1 and 5 take 1.5 minutes/unit to complete their processes; a robot doing task 2 takes 0.5 minutes, task 3 takes 2.5 minutes, and task 4 takes about 2 minutes. Obviously, task 3 and 4 are holding up the whole process. For every unit person 3 completes, a backlog will build up for them. By identifying this bottleneck, you can do what is called *line balancing* (like balancing the workload on a production line) to share out task 3 and 4, if possible, so that all tasks are below the Takt time line (the pace of customer demand).

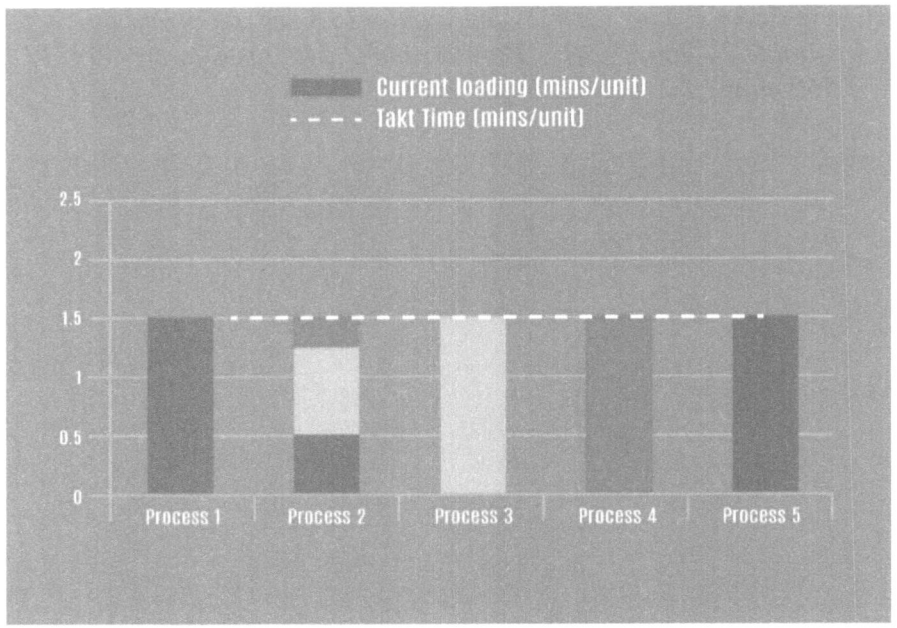

As you scale RPA and start connecting automated processes together from different departments and teams, with human workers interacting up and downstream from your digital workers, this will become a more important consideration in ensuring that your business processes continue to run smoothly.

As an example (Order-to-Cash):

Say that everyday customers are sending in orders for a widget via email. Robot 1 may be responsible for downloading the Excel order and entering the order details into your order system, which takes 1.5 mins per email.

Once an order is entered into the system, robot 2 is responsible for checking the current exchange rate online, searching the employee database for the account manager of this client, and sending them an email with all this information.

Then the human reviews the email, has the widget fulfilled and shipped, and sends these details to the finance team.

The finance team's robot 3 receives order and widget details, reconciles the information so that the delivery matches the order, creates the invoice, and emails this back to the customer.

Let's assume the pace of customer demand is 240 orders a day. The available time for 3 robots and 1 human is 32 hrs. (8 hrs.*4, as these robots only operate during working hours) or 1920 mins.

1920 mins / (240 orders a day)

Takt time (TT) = 8 mins / case. This is the pace in which the business process needs to be working to meet demand and so an analyst will need to review the end-to-end process to ensure both staff and bots have a balanced workload

5 WHYS: Finding the root cause

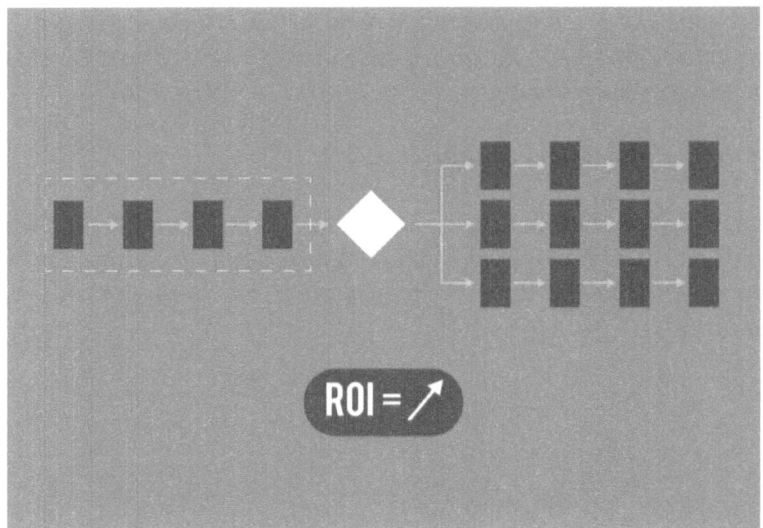

Now that you've identified where the wastes are, you can find and address the root causes. As discussed previously, this can increase your return on investment. There are several ways to carry out root cause analysis; however, in Lean Six Sigma, we're taught about the 5 whys. To demonstrate the 5 whys, here's an example about the Washington monument.

Why is the monument deteriorating?

Because harsh chemicals are frequently used to clean the monument

Why are harsh chemicals needed?

To clean off the large amount of bird droppings on the monument

Why are there a large number of bird droppings on the monument?

Because the large population of spiders in and around the monument are a food source to the local birds

Why is there a large population of spiders in and around the monument?

Because vast swarms of insects, on which the spiders feed, are drawn to the monument at dusk

Why are swarms of insects drawn to the monument at dusk?

Because the lighting of the monument in the evening attracts the local insects

The Solution: Turn the lights on 30 minutes later, so that the city lights attract the insects away from the monument

Such a simple solution for a big problem. You can imagine how teams can over complicate solutions by taking the problem at face value. Similarly, look at the diagram below, many teams may decide to throw lots of automation at the problem where it is being experienced: however, focusing efforts upstream can have the same impact but at a lower cost.

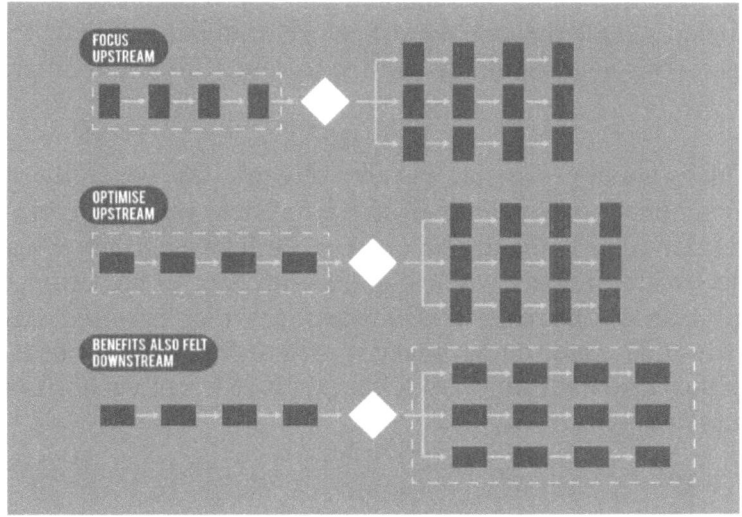

As Is/To Be

A lot of what we've discussed up to now has been looking at the current state of the process, the As-Is. Now we want to look at what the future state or To-Be process should look like once we've reengineered it and implemented our automation solution. As discussed prior, the future process which the bot will perform may be very different to how the human did it purely because the bot is capable of many different things like an exceptional memory and processing at lighting fast speeds.

5 Ss: Organise the physical and digital working environment

Before re-engineering the process, Lean thinking advises that you organize your environment in a way to maximise what you get from your resources and tools. This method can help reduce excess movement of people through the office or through the digital workspace. This also should be applied to your CoE environment too.

Sort: This is about organising the area of work; this generally refers to removing unnecessary items from the physical work area, but why can't this also refer to the digital one too. Perhaps your target team has folders within folders within folders with poor or no naming convention, so no one really knows where anything is stored. And unused files that create clutter make it hard to find what is needed. It could be the case that staff are working with different versions of a form template.

Set (in order): Give everything a specific location, and prioritise materials by the regularity that they are used. The team should have quick access to what they most frequently use, and what they hardly use is stored farther away. It also means if there are materials that have not been used at all, they should be removed from the workspace.

Shine: Once organized, it's up to the team to regularly inspect their environments and keep them tidy. Have controls in place to stop poorly named rogue folders or orphan files from popping up.

Standardize: Define best practices for keeping the environment organized. Create a naming convention and team rules for how materials should be stored.

Sustain: Embed the habit for these 5 Ss into the team so it becomes 2nd nature to sort and keep the workspace tidy.

This is very useful even for your own CoE, as you'll be dealing with an ever-expanding amount of data, code, and possibly sensitive information, so creating an organized foundation will pay off as you scale your operations.

5 Rs: Re-design your process

You could re-design the process from scratch; however, the outcome of this may not be possible to implement. The steps in your process may not actually be the problem, it may just be more about how they are sequenced. Re-using what you already have may be all that's needed to reduce the handling time of a process.

Remove: Remove redundant steps. Some steps may not be needed once the process is automated.

Reduce: The process may need to be reduced in scope.

Replace: Especially with automation, there may now be a new way to complete a step. For example, the bot sends an email instead of a human sending a letter.

Re-order: Changing the sequence could make the process flow smoother.

Redeploy: Move a task to another person or department.

Creating a smoother process flow can remove a lot of wastes, reduce the time the bot spends on the new process, and frees up the bot's capacity.

Are your processes lean?

Before moving to the automating stage, look back and see how much waste you've removed from the process and see if you now have a lean process to automate.

Remember that the purpose of Lean Six Sigma is to meet the customer's expectations as fast as possible, and in RPA and intelligent automation, Lean Six Sigma can greatly increase the ROI of your initiatives. See how well you can answer these questions:

- Have you identified all eight wastes in your target process and team?

- Have you mapped out your target team or process in a SIPOC?

- Have you identified non-value-added areas using a VSM?

- Is your CoE's physical and digital work environment organized using the 5 S method and ready to scale?

- Have you run a root cause workshop and used the 5 whys to discover what the potential root causes are for the inefficiencies in your target team?

- How much waste have you eliminated in the proposed future state model, even before you automate?

Lean thinking is such a powerful combination with RPA, so now you can determine the value they both can bring by doing a cost-benefits analysis and creating a business case. Once signed off, it's time to design the technical solution so that development can begin.

Roll out your automation solution

Now that you've agreed on the solutions to use with key stakeholders and you've planned how to remove the waste in the current process, it's time to implement your automation.

The first step is to create a *business case* to show in detail the benefits vs the costs to get approval to implement these solutions. Once the sponsor/senior leadership team has signed off on this, it's time to plan the solution.

In RPA, the next step after the business case would be to create the *Process Definition Document*. As the name suggests, this is a clear definition of what the process actually does and all the relevant information about the process and the details the developer will need. The previous chapters all laid the groundwork that your analysts have done, and they will now pull all this information about this process into these documents.

Let's look at these now...

22 Business Cases: Cost, Benefits, and Sign Off

The analyst would have created a high-level view of how much each candidate on the automation catalogue would cost to implement—by using 't-shirt sizing' in their cost model, they could calculate an estimated cost by looking at the complexity of the process (be that an XS, S, M, L, XL process). This is what was used to help the senior leadership team prioritise the opportunities. To get real funding behind your projects, your analyst will need to carry out a detailed cost benefits analysis (CBA).

I have actually seen businesses new to RPA and excited by the potential FTE savings decide to mark down the costs as overhead; however, 'turnover is vanity, but profit is sanity'. As not all bots are created equal, it's important to calculate the actual ROI and payback period of each robot. Additionally, that's not just covering the initial expense to build it, but also looking at how profitable the bot will be to run when maintenance and other running costs are considered. There's a share of the infrastructure and the annual vendor licence fees and even server expenses and support costs, so how much does it actually cost to maintain your bots and keep them running?

Hidden charges: What is the real cost of ownership?

The cost of supporting a bot is dependent on a few things: the process's complexity, process metrics such as expected volume and AHT, the bots' run schedule, as well as how well the bot is built. Bot process metrics and run schedules impact running costs, as this will determine how many actual bots are needed.

High-volume processes and processes with high average handling times (AHT) will take a lot longer to complete the batch of cases so will require a larger share of the server, and if the run schedule leaves a small window to complete cases, several bots may need to run cases simultaneously to complete the batch before the deadline.

Bot complexity and build quality impact cost because a more complex bot takes longer for support engineers to analyse to find and fix defects, and a poorly built bot will have more defects and so will require much more support time.

Very rarely is the person supporting your bot the same person who built it, so the support engineer needs to quickly navigate through the code. However, even if it is the same person for a time, it won't be indefinitely. Having code which is poorly commented to explain what each section of code does, or code that is not segmented into different functions, makes it much harder for support engineers to locate defects in code. Also, if the bot uses elements like SQL, C#, or VB scripts, this can increase the bot's complexity.

Other support costs are dependent on the criticality of a process. If it's a Priority 1 process, where defects need to be fixed in a short time frame, this will cost more (even if in wooden dollars for internal clients) if support are to prioritize fixing that bot above other processes.

The support team and IT should provide a breakdown of these unit costs, which the analyst can add into her cost model.

ROI / Payback period

"Show me the money!!" –Cuba Gooding Jr and Tom Cruise in Jerry Maguire

When calculating the financial value of the automation solution, it's important to show how quickly the investment will get paid back, and thus can be re-invested into something else. The faster a project pays off the investment, the less the money is at risk of something going wrong, and the faster the money can be invested in the next project and thus your team can start compounding profits. This gives great confidence for the 'higher-ups' if your CoE team can more quickly show it can 'wash its own face'.

It's also useful to understand what the annual rate of return on an investment will be going forward. This is because you can't compare a bot that saves £50,000 a year to one that saves £15,000 a year. Not much meaning can be derived from looking at definitive values, so for investment comparison, IRR (internal rate of return) or ROI (return on investment) are used to show the potential value generated for every £1 spent on a project.

Illustrating the return on investment for each solution can help prioritize your automation catalogue and guide leadership's decisions. A small quick win may have a much higher ROI rate and much shorter payback period than a larger project, so could be a great choice to start putting results on the board.

Note that not only will this be used to decide whether to invest in one bot project vs another, but also RPA investment opportunities are in competition with other solutions, like outsourcing, lean process engineering, recruitment, or other automated technology solutions. In your business case, it would be good to show that you have considered these other solution options to give confidence that RPA is the best recommendation.

It's all about the money (sometimes)

Including the non-monetary benefits on the business case are very important, as this can sway the decision even if the ROI is less than other projects, or even if the project potentially loses money. For example, the cost of failing to comply with regulations such as GDPR may be difficult to calculate; however, although mentioning that a process which can prevent this from happening may on paper not seem financially viable, the value of this process comes from avoiding fines and reputational damage.

Other non-monetary value may come from an environmental standpoint (e.g. an automated process which uses less paper or has a lesser carbon footprint are such projects where their worth is less explicit and more subjective). This is why the CoE and analysts are there to recommend and guide with data; however, it should be down to the business to make the call in line with corporate objectives.

Some benefits that are difficult to quantify:

- Reduce human error/operational risk (though you could look at the cost of human errors over a period—e.g. how much was lost when an analyst pressed BUY instead of SELL in the last 5 years)

- Ensure compliance with legal or governmental regulations

- Address social, political, environmental, economic concerns

- Enhance customer experience or staff morale

- Embed business practices or guide staff behaviors

- Provide standardization, flexibility, scalability

Business Case

Automation Business case

Process Name	
Business Area	
Analyst	
Process owner	

Problem statement

Please write a summary of the problem that is being experienced, the cost and disadvantages it is causing, and what the situation will be if this is not solved.

Include the team or department name and process name involved

Solution details

Analysis and findings

Recommendation

Impact Analysis

Service Impacts

Systems	Internal or External system
Stakeholder teams (to this process)	upstream, downstream or other

---

The business case should be an objective factual document void of any opinions or assumptions that can't be backed up by data, and should contain all the key information collected and calculated relevant to the process:

- Problem statement
- Summary of analysis and findings, including volume forecasts

- Risks and impact analysis of implementing the solution (including service, systems and teams effected)

- Risk and impact of choosing not to implement the solution

- Cost-Benefits Analysis

  - Intangible benefits and estimated annual cost savings

  - Cost breakdown for initial development and annual running costs

  - Payback period of the recommended RPA solution

  - Return of the investment (up to three-year forecast)

- Other lesser solutions that were considered

- Reference to any relevant supporting documentation

23 Process Definition Document (PDD)

The process definition document is the most important document in the RPA lifecycle. The RPA business analyst uses this to define the To-Be process in unambiguous terms to educate the developer on exactly what the robot will need to do.

Process Definition Document

The purpose of the Process Definition Document (PDD) is to capture the AS IS/TO BE process which will be automated. All supporting documents or links to documents needed to build the robot will be included below.

Once all stakeholders in the Governance table have signed the form and it has been handed to the RPA development team, an assigned developer will start the build

Document details

Version control: (for each revision of the PDD)

Date Issued	Version	Reason for version update	Author
12 Sep 2018	V1.0	First Draft	John Smith

Process Name	
Business Area	
Analyst	
Process owner	
Subject matter expert(s)	

Governance (Stakeholders to sign off document)

authority	Name	Signature	Date
RPA Manager / Technical Lead			
Department Lead			
Team Manager			
Process Owner			

Overview

Problem statement

Please write a summary of the problem that is being experienced, the cost and disadvantages it is causing, and what the situation will be if this is not solved

Include the team or department name and process name involved

---

The PDD should hold all information collected and calculated on the process:

- Process overview

- Stakeholders (subject manager experts, process owners)
- Target systems
    - This should include accessibility levels or access profile types (it's advisable not to give the bot full access to an entire application, just in case. If possible, it is desirable to limit the bot's access to individual screens or areas of an application)
- Impacted business areas
    - Include team SLAs (e.g. team must finish their end-to-end process by 5 pm, so bot must finish no later than 3 pm)
    - Manual workarounds if the bot breaks down
- Risks and impacts of implementation (copy from business case)
- High-level process map
- Key process information
    - Process deadlines (e.g. must finish all 100 cases before 3.00 pm GMT). If dealing with international entities, teams or developers must include time zone (e.g. 'GMT' or 'UK time')
    - Process schedule run times (e.g. must run at 03.00 am and 12.00 pm)
    - Earliest time process can start (i.e. when data will be available for the bot to process)
- Notifications:

- How will bot be notified/triggered to start (e.g. time based 'start at 3 am' or event based 'start when receive email from **abc@123.com**')?
- How should the bot notify the team when it has completed processing cases?
- How should the bot notify team of errors/unprocessed cases (should bot send a report of all errors after completing the batch, or send an email for each error)?

• Keystroke (Level 4) process map (link to or embedded document)

- Having a map with numbering that matches the keystroke document makes it clearer for the developer to understand the flow and decision points.

• Screen recording of process (link to or embedded video)

- Most likely a demo of the process down the 'happy path' (the most straightforward path), with audible commentary on reasons for each decision point
- Difficult to show every different scenario, but this adds further clarity for the developer to see in practice how the process should work

• Keystroke document (link to or embedded document)

- The steps in the process map should be numbered, and they should match the step numbers in this document.
- This document shows descriptions and screenshots of clicks, keystrokes, and decision logics for everything the robot must do.

- Decision logic should also include exceptions and how the bot should handle this (this could be to not process a case but send an alert to the user or to process the exception with different exception handling steps).

- How the bot must handle business exceptions (e.g. if bad data is fed into the process) or system exceptions (i.e. when there is an error with the application or an error in the process).

Development

You may have your development team far from the operations team, perhaps out in the country miles from the city, or somewhere overseas. These developers may never have seen or used the applications before. The business analyst will liaise between the developer and SME. Once the PDD has been handed over to the developer, it will need to be refined, as the developer will be looking at the process from a technical perspective and may have questions that the analyst will need to investigate to update the PDD.

The developer is really good at picking out things like micro-steps your analyst might have missed. Also, as the developer goes through the process to build the bot and navigate through the screens, they will undoubtably call out exceptions not covered.

For example, they may say:

"I've noticed the user copied the username and ID from the document into the form, but what should happen if the user ID is missing?".

Another question may be, *"How long does it generally take for the screen to load?".*

If the screen is known to take a long time to load, the developer may need to programme the bot in a different way. Perhaps the SME needs the bot to wait 30 seconds for page to load, or reload page after waiting 10 seconds. Perhaps the SME would recommend that the bot cancel the transaction for that case and reject it as a system exception and start the process again with the next case. This is why it's important to have regular communications with the SME and developer with BA as translator.

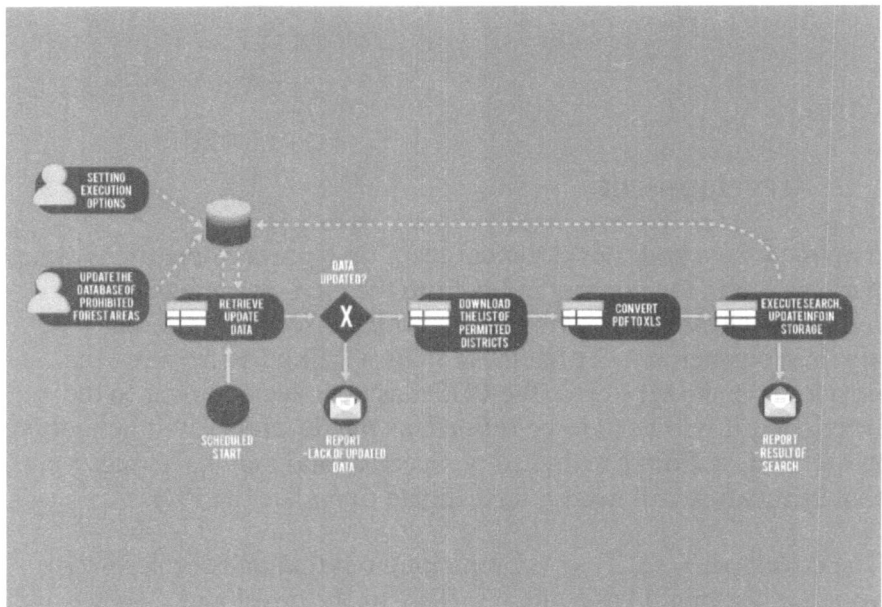

24 Solution Design Document

The SDD should hold all technical information about the process:

- Solution overview

- Solution diagram (including interaction with databases and systems)

- Solution description

- Object model diagram (business process, system objects, utility objects)

- Operational control and alerting

  - Scheduling or starting – what will trigger the bot to begin?

  - Alerts – how to notify user of completion or of errors. This includes how the user wants to receive alerts (email, report) as well as the type of information they want to receive and by when (e.g. users need to know before 1 pm of any errors so team has enough time to manually process rejected cases).

  - This should be in table form showing scenario, method, and recipients.

- Data security and credentials

  - Information on data storage (where files will be stored or database location), data privacy (secure locations or eyes-only accesses), and data preservation (e.g. backup data)

  - The regulatory methods and checks for handling data so bot is compliant (GDPR, CASS)

  - The type of credentials/access levels the bot has for each application

- Technical and business assumptions

25 Best practices and standards

Working inside a Support CoE, assisting in the management of support engineers as well as release management of changes to existing bots (be that to fix defects or make enhancements), I had a unique insight to the issues caused by poorly designed RPA bots.

Rushed or badly built bots end up costing a lot more in the long run. End users may not stick to the strict guidelines of how to interact with the bot, or as staff change roles, new staff may not have even received the memo. Your bots need to be able to handle every eventuality as a first line of defense; failing that, your support team need to know how to fix it quickly.

From firsthand experience, I must express the importance of ensuring that code is commented and segmented in a way that makes it easy for someone other than the developer to quickly navigate the code. It is not uncommon that a developer contractor is hired, builds the bot, and then moves on. In these situations, they're not around long enough for when the bot falls over, after it inevitably digests a bit of data it wasn't prepared for.

Your solutions architect will need to create coding standards and best practices for all developers and support engineers in your organization to follow. These standards much be framed with longevity of the bot in mind.

Without getting overly technical, the main areas to consider are

- **Documentation** – Adequate knowledge transfer so any developer can understand how to modify or support the bot.

- **Modular code / reusability** – Segment the bot into components for steps (e.g. log in/out, data validation, etc.), this makes it clear in the logs where the issue happened.

- **Error handling and logging** – The bot notes the outcomes after completing each step in the process. If there's an error, it can reference in which component the error occurred.

- **Commenting/readability** – Well-laid-out code so the logic is clear, and well commented so someone else can read what each section does, and why certain logic was used.

- **Variables and referencing** – Variables represents some value which can be referenced in several places in the code; however, the developer only needs to make the change to the variable to ensure that the value has changed everywhere.

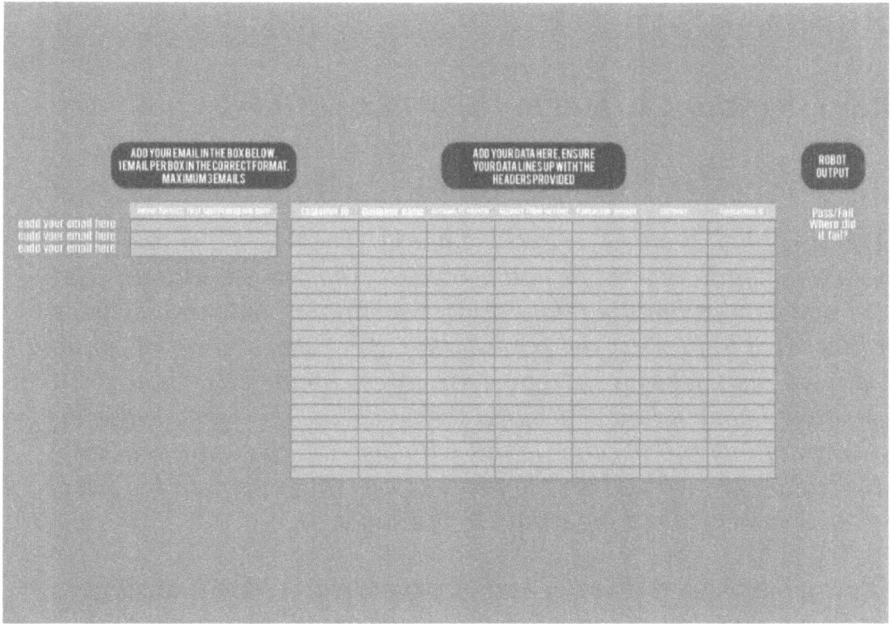

- **Test planning** – Having a structured approach to testing saves time. Testing of each component individually, then as a whole, is easier to fix than building and testing the whole thing at the end and then having to look through the whole process to find the issue. Also, it's advisable to test to failure rather than test to success. The tester (preferably a developer other than the original builder) should try to break the bot, using various scenarios and bad data. This should be done even before UAT.

- **User flexibility** – This is designing the robot in a way such that the user can make small changes to the process without the need for the developer—e.g. creating an Excel spreadsheet and data table which could be updated for the robot to process new information, as shown in image above or allowing for several spreadsheets to come from different sources and still all be processed by the same bot.

REAL-WORLD EXAMPLE: USER FLEXIBILITY

As a BA, I had created an excel template for the SME to add their data and email into the format provided. The process started with just one team submitting data as an Excel file to the robot's shared folder for it to process the information. After this, the robot emailed the team representative a report for cases completed or not completed once it had processed the entire list. However, after the success of this robot, other teams who also did the same process wanted to get involved, so the process was then expanded to have several teams submit their data to the robot.

During the original build, I had discussed this with my developer, and we built into the robot for it to check for all files in that folder with today's date on it, and then using the emails added to the template, send the email confirmation and report to those recipients.

As a result, a little forward thinking resulted in over a thousand more FTE hours having been saved each year from the same robot without the need for the developer to make any adjustments to the code. Hence, the benefits realized were more than was originally expected.

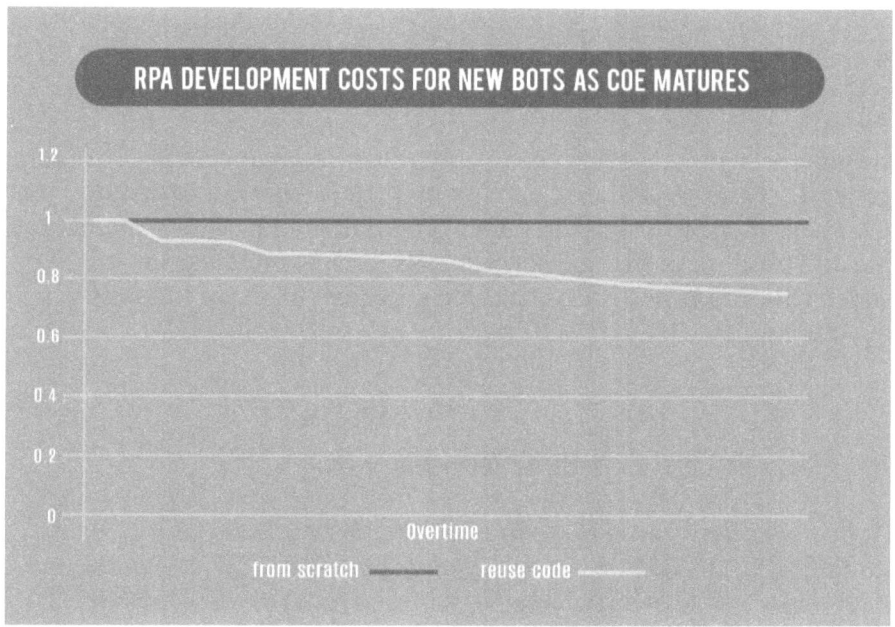

Reusing code and creating libraries

We mentioned about modular coding (or modular programming), and this related to the bot economy mentioned in an earlier chapter. As your developer team build more bots, they will start to notice that many tasks inside the automated process are the same as projects they've worked on before, as processes within your organization are executed on the same applications and used in the same way. Your development lead or solutions architect should encourage your developers to store reusable components in libraries for other developers to effectively copy and paste code into their bots.

As discussed in the Bot Economy section, similar businesses in the same market sectors and similar functional departments such has HR and finance also tend to use the same applications and use them in the same way. For example, logging in to SAP or adding an employee or managing payroll is a standardized process and would generally be done in the exact way no matter what company you work for. Your solutions architect should be looking into whether it makes sense to acquire components or perhaps purchase larger standardized processes from trusted suppliers to shave time off development.

REAL-WORLD EXAMPLE: Reusable robot

After being assigned to automate a specific process where I identified potential savings of £150k from 2018 to 2020 (based on the company's growth forecasts), with the help of management we took a zoomed-out view of their whole department, to discover that another team based in a different city had almost the exact same process of aligning data between the same two legacy systems. Working closely with the developer, we identified that we were able to re-use most of the robot, as the only real different between the two teams was that they were processing customer details of who had purchased different services, so the team were carrying out the same steps on the same screens but where selecting different dropdown options in the application's forms

26 Data preparation and test cases

Whilst the developer has been building the bot, your Business Analyst should have been busy working with the SME to gather the dataset needed to test that the robot can process it in the right way. You may need to anonymize the data first. However, collecting a set of data from previous transactions is significantly better than making this up, as fake data may not truly reflect reality.

To know what data you need to collect, you first need to understand what you need to test. The SME should assist with creating the test cases or scenarios they want the bot to go through, different data types, different exceptions, and various known issues. This is just to test that it works well. Then it's important to create bad data, wrong formats, and missing information to try and break the bot.

If, for example, a field on the input form is a dropdown field but the SME says that the user will only ever select 'YES' or 'NO', the business analyst needs to be able to ask the question, 'but what *if* they left it blank?'. Many times, if the user has never seen something happen, e.g. they always see YES or NO for the 5 years they've worked there, they will assume that it can't happen any other way. The business analyst needs to think laterally about anomalies, as these will also need to be tested.

This is because, if the input form doesn't have the right conditions, it may be possible for a user to leave a field blank, or if it's a third party's form, they could change it without giving you notice. If the developer is lazy in their code and writes logic for only the known outcomes: *"IF YES do X, ELSE do Y"*, instead of *"IF YES do X, IF "NO" do Y, ELSE log error"*, then they've set the robot up to fail if sometime down the line the process changed to allow something other than YES or NO, like 'maybe' or it's left blank.

The test data provided to the developer to use when she tests her robot during build stage should be a completely different dataset to the dataset the SME and BA will use in the UAT, as it's very easy for the developer to unintentionally manipulate the code until it works for her dataset specifically.

27 Testing: user acceptance testing, debug, sign off

The developer will need a handful of test cases as they finesse their build before it is ready for UAT testing. The developer needs to know that the bot is behaving properly and that it's catching errors and writing logs at the right points throughout the process. Once the developer is confident the robot is ready, the bot is run in the test environment for the user to interact with and then inspect the outputs of the robot.

Use Lean Six Sigma's SIPOC to test every part of the process.

Supply: Where did the information come from?

What if the file is missing from the location the robot gets the data from or if the email sent to the bot doesn't contain a file?

Input: What is the type of data?

What if data is missing or in the wrong format: e.g. the user copied the email <'first.last@company.com'>, instead of just first.last@company.com without the parentheses) or the file is the wrong format (word instead of PDF)?

Process: What are the process steps and logic? Check the positive and negative possibilities for each step.

What if the site page didn't load, what if the credentials are wrong/expired, how should the bot handle errors and exceptions?

Outputs: What is the type of data?

Are the outputs correct and did the bot handle errors and log these in the right way—e.g. does the application in the test environment show that the bot made the transaction correctly (correct accounts, correct direction, correct amount)?

Customers: Where does the information go?

Were the outputs of the process sent to the right people—e.g. did the bot send the notifications and reports to the right staff?

After the process owner is satisfied with the results of testing in the test environment and the bot has gone through the correct release management process of approvals and handed this to the support team, it's time to test in the live environment.

It's important to expect the bot to behave differently than in test and be ready for this. Certain high-impact processes may require testing for a high volume of cases or for a longer period in the test environment before the bot is deployed into production. Even once it is deployed into the live environment, it is advisable to only direct a small volume or, if financial, low-value cases through the bot, and test this for a couple of weeks whilst gradually increasing these thresholds.

REAL-WORLD EXAMPLE: UAT

Input	Expected results or Output	Output Metrics
Correct credentials	Correct (user logs in)	Passed
Incorrect credentials	Incorrect (login fails)	Failed
Correct link	Correct (ACME System site is opened)	Passed
Incorrect link	Incorrect (ACME System site is not reachable)	Failed
Correct (normal browser scale)	Correct (all web actions are performed)	Passed
Incorrect (changed browser scale)	Correct (all web actions are performed)	Passed
Correct value for filtering table	Correct (table is filtered)	Passed
Incorrect value for filtering table	Incorrect (table is filtered but next actions fails)	Failed
Correct excel table path	Correct (all excel actions are performed)	Passed
Incorrect excel table path	Incorrect (table has wrong format)	Failed

To minimize disruption of the BAU team, once the developer had completed their build, I as the business analyst, interacted with the bot using a sample set. An Excel file of test data was placed into the shared folder and accessed by the bot. It digested the information and created a new file with the processed data saved in the 'Completed' folder. Once completed, the bot emailed me that it had finished and whether it had successfully processed all cases or if there were any errors.

Using a set of test cases or questions created by the SME and my understanding of the process obtained from the 'Design' phase, I was able to provide feedback to the developer on certain misunderstandings or incorrect behaviors of the bot for him to edit the code, so that we could test again on a new set of data. This was repeated, involving the SME where needed, until the robot was 100% successful at handling each scenario (and by that, I mean it also handled errors and cases it couldn't process in the anticipated way). Only then I had the SME get involved to interact with the bot and verify the outputs of the tests.

28 From live testing to support BAU

During the live test phase, this may require all hands-on deck with support engineer, developer, and SME all on-call to babysit the bot should something go wrong. The support engineer should use this period to learn as much about the bot as possible whilst the developer and SME are still fully engaged. Before the Go-Live date, they would have received all the documentation, picked the brain of the developer, and probed on anything that they were not clear on. After live testing is complete, this should be a smooth transfer of the bot from the development team to the support team.

Just as the development team have their own criteria for receiving documentation from the BA, the support team will have criteria that the developer must meet, including how successful the robot was during testing. The support team have finite resource, so they need to best predict how much time is needed to support the new bot.

Similar to how complexity of the process was used for the developer to calculate the time needed to build the bot, a complexity calculator would be required to estimate the time per day or week to dedicate to supporting this bot.

An added element to consider when supporting technology is priority levels. The support team is a finite number, and one support engineer can manage about 7–10 medium complexity robots each. However, if multiple robots go down at the same time, not all incidents can be handled at once.

How quickly after the robot goes down will the business be impacted? Straight away (i.e. wrong data sent to clients)?. Or will it take a few days until the backlog of work becomes unmanageable? Questions like these will help determine the priority level and thus how quickly a robot needs to be fixed.

29 Launch

Celebrate your success. Get the word out

30 Reflect on your achievements

"Those who don't fail don't try"

Your pilot was successful, you've implemented your first handful of bots using best practices from your CoE framework, and the momentum is up. Don't be afraid to admit you've had some hiccups, as this is only natural with anything new and means your team have learnt a lot. Make sure you're capturing the lessons from each project so that for each future project your team gets better and smarter.

Make sure you are continuously reviewing lessons learnt and if possible fixing the problem as it happens, otherwise these will be more like lessons *'experienced'* rather than lessons *learnt*, which is generally the case in many organisations.

It's good practice to compile a lessons log (your CoE should manage this as part of their knowledge management) so that for any new project, lessons from similar projects can be brought up and pitfalls can be avoided. Remember: "A person who never made a mistake never tried anything new." - Albert Einstein

Let's take a recap of the chapters so far:

- Aware: Are you and your team aware of what's happening in the RPA world?

- Aligned: Do senior stakeholders agree on the purpose of the Automation CoE team?

- Educate: Have external experts educated your CoE and those involved and interested in RPA?

- Empower: Have you empowered staff with skills, tools, and techniques for identifying good RPA opportunities?

- Inspect: Have you mapped out RPA candidates on a cost-benefits chart, created an Automation Catalogue of prioritized opportunities, and created an Enterprise Automation Roadmap?

- Ideate: Have you combined the experience of subject matter experts to design lean solutions to solve root causes?

- Optimise: Have you applied lean thinking to remove wastes before implementing your automation solution? And have your team used the right standards for building and testing your bots?

31 Scaling up: RPA factory (repeatable)

Now that you've been worked through the first handful of processes on your automation catalogue, you should have tailored this framework for your organisation and created an approach that's repeatable:

Identify candidates > assess suitability > measure > focus > assess complexity > prioritize > create business case > create PDD > Design > Build > Test > Support

Now you can keep that drum beat going and start churning out more new bots.

- Go back to your prioritized list of opportunities in step 17. Follow steps 18 – 29, launch your next set of bots, and repeat.

    o Simultaneously, your support team must be able to keep up with the new bots coming in and have enough foresight to hire and train staff in time.

- Keep communicating your successes so that your pipeline stays full and you keep momentum going.

- Systematically refine your tools, templates, and criteria at each stage. Before long, you will get served a curveball you were not prepared for; this will help you to provide a better, more robust service next time.

- As your CoE evolves, the structure may likely change. Clearly defined up-to-date roles and responsibilities are critical so everyone knows what to expect from each other, and what is expected of them.

- Every project is an opportunity to learn. Keep your lessons learnt repository alive and keep refining your framework.

- Elevate the maturity of your CoE by developing better governance and controls. A programme management officer can help maintain standards, maintain your knowledge base, and track performance of your team and your bots to continually identify new places to improve. They can also help your team identify risks and issues.

Visit www.leania.co/bookimages and click on the RPA team test, answer multiple choice questions to receive a comprehensive personalized 11 page report, identifying your teams strengths and weaknesses and CoE maturity for each part of the AEIO YOU framework

GUEST: RPA public speaker and author

*I reached out to **public speaker and author Rob King,** who is the **co-founding director of management consultancy firm Wzard Innovation,** to talk about his experiences of the industry and challenges that businesses have with scaling.*

Rob, please share a little bit about your experience in RPA and intelligent automation.

My career started in IT where I was developer, analyst, and IT manager before jumping over into the business side where I led numerous lean and Six Sigma programmes.

My first experience of RPA was during a lean transformation programme I was running in a life and pensions business. Opportunities for automation are a key element of transformation, and there was a natural fit.

Even in 2012 the desktop product was remarkably mature and effective. Our first implementation rolled out to 47 desktop machines.

The year 2012 was a turning point as I also took responsibility for an innovation initiative (called "digitise the enterprise", which fits nicely with RPA). RPA and low-code development were the two most important tools in the toolkit.

I left corporate life in 2017 to set up a management consultancy, and my experience in RPA from a hybrid perspective, both technical and business, was a key element of our USP.

Personally, the effective use of technology to accelerate change is what attracts me to the industry, but I still continue to work outside of RPA also, driving innovation and change.

What trends have you noticed in this space?

This is a tough question, because the industry continues to surprise me. I'm sure my imagination is simply not enough. But I will take a stab.

The advances in AI, or machine learning (ML) to be more specific, have equally been surprising in the last couple of years. Elements of ML have become a commodity, opening new opportunities for businesses that would have required deep pockets only a year ago.

However, alone, this commoditisation of AI isn't helpful unless there is a way to glue it all together. And there are realistically two very good ways to do this. The first is RPA, the second is low-code. RPA will continue to be the glue that links up AI capabilities to provide the intelligent process automation that is needed by businesses.

The trends to make intelligent automation easier to use, adapt, and support will continue to advance.

The 2nd trend I've seen is, a move from "do it yourself" to looking for partnerships is increasing. While it is cost effective to do it yourself, it's also a big headache if you don't have (or can't hold on to) the right resources, partnerships, or outsourcing to smooth out some of these resourcing problems.

Also, many companies ignored the need for support and maintenance from their business case, so the running costs were unexpected. The level of awareness that this is an important element of planning has increased, thus also pushing more businesses to seek partnerships or outsourcing arrangements.

Finally, the realisation that there is more unstructured data than structured in organisations, through lengthy emails, documents, and other forms is increasing the need for machine learning. While many 3rd party companies have been operating in this space for a while, the main providers now see this as an essential component of their platform and will start to incorporate this functionality, making it a one-stop-shop. Automation Anywhere's iQBot is just the first example, the others will follow hard over the next year, and 3rd parties will have a more difficult time surviving (that's my crystal ball-gazing prediction).

I think UiPath made some very clever strategic decisions two years ago (the free training and the community edition). They are reaping the rewards of this now with lots of practitioners and small businesses who have grown up with UiPath. As these individuals move through the workforce, it will become easier to find individuals and consultancies with UiPath experience. Combine this with a market valuation of $7 billion, and they're becoming an unstoppable force.

You talked about AI, how will cloud impact the RPA/intelligent automation market as well?

Cloud is really a hygiene factor these days. NOT having cloud will be detrimental, rather than having cloud being beneficial. I hope that makes sense.

The value is huge though, cloud is reducing IT cost, increasing (mostly) system resilience and uptime, and eliminating the need for costly strategic investment in hardware. It's the move from capex to opex which cloud has facilitated, and automation benefits from this value-add like all the other SaaS (software as a service) services.

Furthermore, from a process perspective, many processes have seasonal demand (month-end / year-end / etc.), and traditional licencing does not fit businesses that have this problem. Cloud provides the flexibility to increase/decrease utilisation to need and solves this business problem. Though licencing still needs to catch up with a better utilisation-based model, it can still be cheaper to licence old-school at the moment, so a good math brain is needed at the moment. That won't last: as the experience of cloud-based automation increases and demand increases, one expects costs will fall.

What hurdles do companies have with implementing RPA?

The two biggest problems are arguably "pre-RPA", and those are lack of action due to a lack of understanding of what RPA can do (for companies of any size), or awful misconceptions about size, cost, or complexity that is putting people off!

The culture and mindset that best fits RPA is that of a "Citizen Developer", but this can be a challenge if the environment doesn't support it. And IT can become a roadblock if they are not correctly engaged. It's a different way of working that most organisations (particularly of scale) have any experience with.

.., and how about when companies try to scale?

Once you begin looking to scale up, you can run into your first challenge if you first jumped on the bandwagon without thinking which solution fits your requirements best. While some solutions have similarities, others have strengths in key areas that should be matched up against business needs. Early adopters who are running a second solution in parallel are examples of companies that lept before they looked. Companies have struggled to scale because of a poor choice of solution.

The skills gap comes up in every conversation at the moment. If you choose a solution poorly supported by training, partners, or potential employees, then it will probably only really hit you when you look to scale.

The actual activity of support and maintenance (and by definition the associated governance) is often more complex than first thought. The number of variables to consider, to operate at a scale beyond the initial pilot projects, are significant and hindered by a lack of experienced individuals.

How do you see RPA and IA evolving?

RPA and intelligent automation will just be a tool in common use—the Microsoft Windows of this generation. No one really struggles to find their way around a Windows PC, and there are alternative operating systems that fit niche roles. Intelligent automation for a while will be the same, it's a tool and every business will have one.

The more advanced technological companies that perhaps bypass RPA in their early phases as core systems and provide straight-through processing by default will gain back office systems and support functions that will benefit from RPA, so no company is really going to not-require intelligent automation.

Continuous Improvement

Digital transformation isn't an event or a one-and-done thing, it's a continuous process of enhancing your business, reviewing your enhancements, and looking at how it could be better. Once you've automated all your RPA-suitable processes, you still have a world of opportunities that AI can solve. Remember, RPA can only understand 20% of your data, which is structured.

Incorporating new automation and AI capabilities will open your eyes to new insights, new avenues for your business to venture down, and new ways to approach problems or gain more value from your resources.

This completes the end of the AEIO section, bringing RPA and intelligent automation into your organization. Now that you've implemented it, it's down to **YOU** to evidence the benefits and sustain them.

YOU

The AEIO stages of our methodology have taken you all around the RPA (robotic process automation) lifecycle, from idea to implementation to intelligent automation, and you've handed your bots over to BAU.

We've gone step by step to ensure that you've approached RPA in a logical manner to get the most out of this new and fast-evolving technology. We are hopeful that your business and its operations teams will become much more efficient now, and staff across your organization will really start to feel the improvements at an individual level, too. If you're in a large company, senior leaders in your organization will notice how much more agile teams work as they are able to make better and faster decisions. If you're in a smaller company, you'll realize how much potential you have now, powered by an instantly scalable workforce allowing you to make much bigger moves in your marketplace. But guess what...

Launching the bots is only the beginning!

Now that you have bots in production and your design team are busy working on new RPA candidates, your CoE need to be thinking about three questions:

1. Are your bots actually delivering the expected benefits?

2. Are you in control of your bots to ensure you can sustain the benefits realized?

3. How can your bots be enhanced to deliver more value?

Coming from a finance and continuous improvement background, it's always important to look back and measure, did the investment work as expected? As your CoE team manage the bots, you have access to the bots' activity. So your clients (internal or external) will want to know how much their bots are actually saving them. This benefits you two-fold.

- You can tell a powerful success story on what you've *actually* delivered clients.

- Clients can use this same success story to secure further funding for more bots.

Y: Yield

You've measured the old way. Now, the new way needs to be measured and compared, so that we can confirm that your predicted benefits have been realized. How close were you to what you forecasted on the business case? What were the reasons for the difference? This is the most important step, as this is the whole reason for the change.

Do note though, financial benefit or FTE savings isn't the only yield we should be concerned about. Error reduction and improved customer (or user) satisfaction are other metrics to measure. However, by definition these intangible benefits are difficult to quantify, though not impossible. Even such things as human errors can have a financial value tied to them.

For human errors, you could look at the total losses in the process which were due to mistakes made in a specific time period (e.g. how much was lost in the last year of staff entering 'Buy' then reversing mistake to 'Sell'), or you could look at the manhours taken to reverse errors made by staff.

For customer satisfaction, you could use net promoter scores (NPS) to see how clients rated your service before and after the change, though you may not easily be able to determine what improvement of the score to attribute to your new RPA capability. Perhaps you could look at customer attrition numbers before and after and looking at their reasons for leaving (be that due to delayed, slow or inaccurate service) you can consider the cost to acquire a new customer.

It's not entirely true that qualitative data can't be quantified. There just needs to be a logical and consistent way to measure improvements.

32 Realising the benefits

Whichever metric or performance indicator you are using RPA to improve, putting little effort or finger-in-the-air approaches to measuring performance beforehand and a lax approach to highlighting all the intangible benefits can lead to initiatives not capturing the full extent of how much RPA has really benefited a business, and results may seem to give sub-par returns on investment.

As mentioned in the lean thinking chapter, measurement is one of the six areas which can cause errors. The accuracy and method by which you measure are as important as the efficiency of the process you are measuring.

Tangibles & Intangibles, Qualitative v Quantitative

As previously mentioned, FTE savings has been the main focal point in the RPA space for the last few years. However, RPA is much more than that. Some RPA CoEs may have a financial saving threshold so that only highly profitable bots are considered. This makes sense, as some processes aren't financially viable to automate and are not worth the time spent. However, compliance processes like GDPR (General Data Protection Regulation) or CASS (Protection of Client Assets and Money), where speed and accuracy are vital, these processes just needed to get done. In legal and compliance situations the tangible benefit is the cost avoidance of not being fined.

Measure the right metrics

The following are some performance measures to key performance indicators (KPIs) for your bots and your CoE teams. KPIs should always be used to answer a question on how to improve the quality, speed, or costs of your CoE, or to provide insights to clients and senior leadership.

Bots:

This is to determine how well your bot has been built, and how efficient your bot is.

- Actual FTE time saved or annual ROI – *Are financial benefits being realized, and how could this be improved. Perhaps the underlying process can be re-engineered if this has not been done already?*

- % of cases successfully processed – *Has human error rate improved?*

- % of cases resulting in systems exceptions – *how badly has the bot been built?*

- % of cases resulting in business exceptions – *how flawed is the underlying process?*

- Bot utilization – how effectively are you using your robots?

Determine how efficient your CoE team are at building and maintaining your digital workforce.

Design team:

- Amount of revisions to PDD/SDD – *How detailed are your team making documentation the first time around?*

- Build time for each complexity level – *Are developers coding more efficiently, by re-using code etc.?*

- Estimated build time vs actual – *how accurately are your developers estimating build times?*

- How fast are bots launched once the build is complete – *Are there any blockers or bottlenecks in the test and deploy stage, or perhaps there's excess bureaucracy that's stalling the flow of your RPA factory?*

Support team:

- Estimated defect fix time vs actual – *How accurately are engineers estimating fix time?*

- % of SLAs met – *Is your support team on top of defects? Otherwise, what are breached SLAs costing the business?*

- Fix time for each complexity level – *Are support team fixing faster by re-using code etc. Does this vary from person to person?*

- Support time for each complexity level – *Is support team maintaining bots better over time, by optimising their processes, perhaps using automated checks? (Yes, bots checking bots)*

Enhance for better ROI

You can improve ROI by reducing the average handling time (AHT) in which the process is completed or increasing the volume. Implementing Lean Six Sigma techniques can reduce the AHT by making the process more streamlined. As shown in an earlier case study, I re-designed a process and used the same bot for several other teams, so the volume it processed increased without the need for more development.

Reducing errors is also another way to save time and increase volume. Robots may not waste time re-trying failed cases but reject that case, so the robot will process more cases if less errors occur. Two ways to improve errors are to reduce *systems exceptions* and reduce *business exceptions*. Systems exceptions can be reduced by ensuring that bots are built using your best practices. However, if the original bot was built before these practices were enforced or built by someone else, then your developers may need to re-build parts of, or all of, the bot.

To reduce business exceptions, you need to improve the underlying process upstream by improving data integrity. Is the bot being feed bad data, and is there a way to ensure only the right data is sent to the bot using data validation methods?

Train your staff

Perhaps you're receiving errors and exceptions in the process because the end user is adding the wrong data format in the field. The seemingly quickest and easiest way to rectify this is to make the staff aware of their error and train them on the correct format. I did say seemingly, mind, as some staff may forget or be lazy, or new staff may not be made aware of the new process.

So to ensure that these changes are sustainable, the analyst who engages with the RPA team should ensure that the team manager revises their work instructions and the team understand not just the new process, but understand the implications, as a broken bot would just mean that they would need to manually do that work thus creating more work for themselves.

Improve the front end

Though not necessarily achievable immediately due to time or budgetary constraints, a more permanent and reliable fix for data integrity is to put conditions on the front end. That is re-designing the input form so that it takes dropdowns or radio button selectors rather than free text, or it checks for the correct format of an email, xya@companyname.co.uk.

And, if you're thinking, "what if we don't have the ability to change the front end because it's a 3rd party site?" In this case, you will need to build the conditions into the bot itself.

NB: As your developers should know, best practices suggest only accepting positives and throwing out the rest for manual processing:

i.e. have your bot say IF status = 'Married' then do X, IF status = 'Single' then do Y, ELSE pass to user, instead of: IF 'Married' do X, ELSE do Y.

The reason being, if this is a highly sensitive matter (like insurance or finance), if the underlying system or input data changes down the line, you could be sending the wrong information to clients, which could risk your business reputation or end up in fines.

Are you getting value?

The big question clients, stakeholders, and senior leadership will ask is, are they getting value from this digital workforce? Look at these questions to make sure your team can answer this:

- Was performance of the processes measured accurately before they were automated?

- Do you have visibility of your bots' performance, via logs and a dashboard?

- Do your internal/external clients have visibility on how their bots are performing?

- Are you regularly tracking the bots' and CoE's performance metrics that answer the right questions?

- Have you found ways to quantify qualitative data?

- Have your robot performance indicators helped teams discover potential ways to improve on the most common systems and business exceptions?

O: Organise & Oversee

Once the POC(s) have been successful and handed over to a support engineer to manage, a governance process can be established which best suits the environment and business. This means setting up a Support hub inside your Centre of Excellence team to keep the new robots running and be able to quickly fix any defects.

At this stage, you want to create standard processes and embed controls to smoothly transfer bots from your development team to your support team, using the lessons you've learnt throughout the pilot process. The Development and Support teams operate in very different ways; however, they should be seen as two halves to your Centre of Excellence. The development side is dynamic and project oriented, whereas the Support team is generally repetitive and predictable, for the most part, when supporting well-built bots. Only when incidents happen or process owners request enhancements due to a UI or process change will the support team spring to life.

The Support CoE team

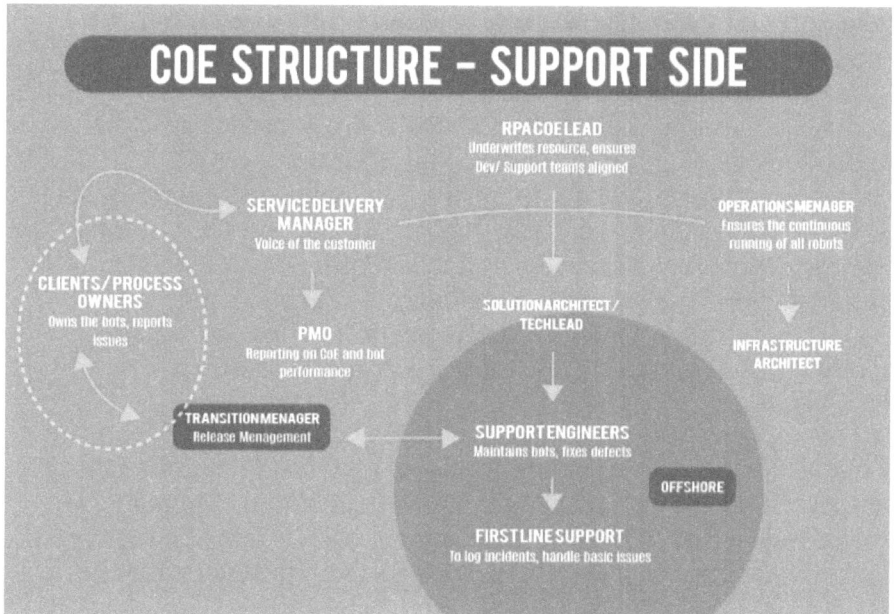

Your new Support side of your CoE is responsible for providing templates to add to the CoE's repository (criteria, checklists, templates etc.). They are the best placed to design and refine procedures for managing new bots, managing incidents, and changes. The **Operations Manager** is accountable for capacity and capability management, as well as designing the disaster recovery and business continuity plans, as it is her responsibility to ensure that the robots keep running, even in the event of a power cut or internet failure.

Furthermore, the **Transition Manager** is responsible for protecting the live environment and so must have his GRC (governance, control, and risk) toolkit to ensure every new release goes through the correct governance, meets all the criteria, and all engineers follow the right procedure to avoid wrong code or untested code being released in error, or at the wrong time.

The **Service Delivery Manager** faces the client or process owner to maintain a good relationship through the management of queries and concerns and the regular reporting of bot health and performance. This may include reviewing the business exceptions as well as systems exceptions to provide advice on how to improve them—perhaps a re-build or modification of the bot to reduce systems exceptions, or suggestion for re-design of the process or enhancement to data integrity upstream. Comparable to the Business Analyst in the Design team, the Service Manager with the help of the Support Engineers has access to the data of the bot and the automated process, so has the visibility to see where potential opportunities of continuous improvement lie. This data can be provided back to the RPA or Lean Analyst to do further investigation.

To assist the Service Delivery Manager and the Operations Manager with tracking and reporting on the CoE team and the digital workforce, a **PMO** (project management officer) may be required to gather and generate insightful reports for both the client and the **RPA CoE Lead/Sponsor.**

As you can see from the Support structure diagram, only the Service Delivery Manager (for business relationship management) and the Transition Manager (from a release management perspective) interact with the client or process owners. This way, the client only has one point of contact for each need, and this scalable structure allows for more managers to come into the team, as your CoE customer base grows.

This is opposed to different clients or process owners speaking with different people from the support team, which could be a different engineer every time, this would be difficult to manage and messy to scale. It is also important to have experienced senior people controlling communications between customers and the support team.

Nearshore? Offshore? …'Cybershore'?

Not only have I experienced many RPA development teams having their developers based overseas with the Tech Lead based in this country, but I've also seen Support Engineers based offshore too, or at least first-line support. This has many obvious pros and cons.

Working in an Operational Excellence team whose developers were in the US and Support team were based in India did pose several difficulties, such as time difference and language barriers, and some say that the United States and Great Britain are two countries separated by a common language. However, having some engineers and developers overseas would allow for 24-hour support, which is vital if some of your bots run overnight, and there is also a lower support cost.

Taking these cost savings one step further, there have been discussions in industry about having certain support tasks such as first-line support and regular bot checking being carried out by other robots. Bots managing bots may sound a little frightening, and many processes such as finance will require a human in the loop. However, it is ironic to not want to automate an automatable process, just because it's in the Automation team.

Design – Support team interaction

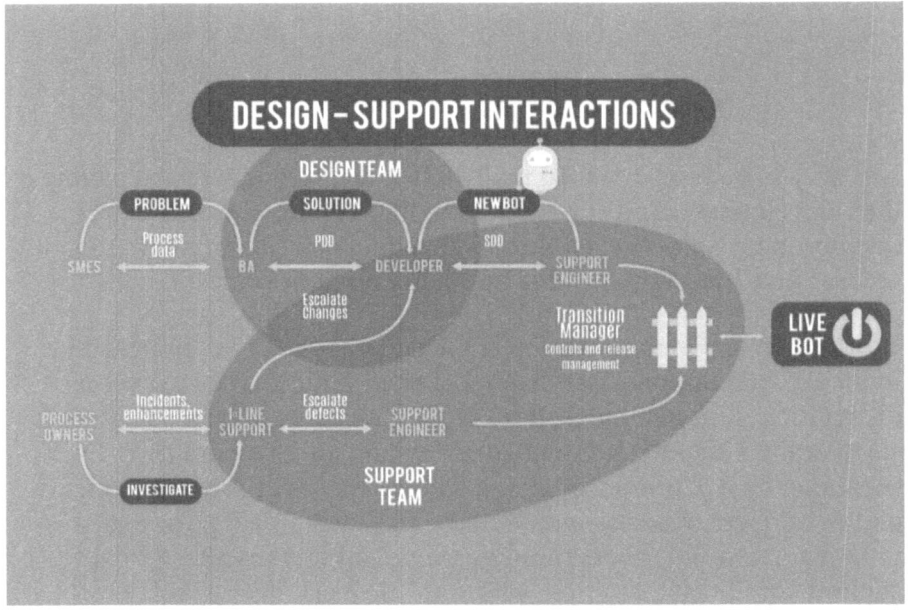

On the previous image is a simplified diagram depicting how the design and support teams interact with each other. As you can see, it is the Support team which is the iceberg below the surface. **SMEs and BAs** work together using process data to define the problem; the **BA and Developer** have a close relationship to produce the PDD (Process Definition Document) in order to define the solution showing exactly what the bot will do. The Developer will design all aspects of the bot using the SDD (Solution Design Document), and once built, the **Developer will transfer their knowledge to the Support Engineer**, and going forward the SDD will be used to manage the new bot.

Even though the developer owns the SDD documents, it's important that the Support Engineer has some involvement in refining the document. The SDD should be detailed enough that the Support Engineer fully understands the process. However, the SDD is a living document, so as new issues, exceptions, and modifications arise, the Support Engineer should update the SDD so that any new engineer can take over.

33 Maintain your benefits

Keeping the bot running, keeping exceptions down, and minimizing downtime is what the Support team need to do to ensure that the benefits the BA forecasted on the business case are actualized.

The Design team get the glory for delivering, but it's actually the Support team who deliver the savings.

The engineers, or perhaps 1st line support, regularly check to make sure the bot is up and running, but also if it is executing the process correctly.

Teams, clients, bots, and processes all differ and may require being handled in different ways; however, the underlying method should be the same. SIPOC is a good guideline for what to include in your standardized checklist when reviewing your bots. As mentioned earlier in this book, SIPOC stands for Supplier, Input, Process, Output, Customer.

S (Supplier): Where did the data for this process come from and how was it received?

What is the file location, what specific person/mailbox should the email come from, what specific query should be given to the database? Was it sent by email or from a website chat?

I (Input): What does the data look like?

Should it be in a PDF or Excel file? Should all fields in the database record be filled in? Should the address field be a city or a country?

P (Process): What steps should the bot have taken?

Did the robot make all the right steps per the process map? And did it do this in the right order?

If hundreds of cases were processed, check the logs of every xth case (e.g. every 10th case) against the current test case questions. Did it pass?

O (Output): What does the transformed data look like?

Should field XYZ be either Yes or No, or can it say 'maybe'? Should the address field be a city or a country? Should it be in a PDF or Excel file?

C (Customer): Where did the data go and how was it sent?

As you can see in the above checklist, all of these steps are logical and routine, as it is just a check to whether something matches set criteria. Furthermore, these checks in themselves are susceptible to human error, and the human would not be able to check hundreds or thousands of cases for errors. Unlike a human a bot is capable of checking all cases, accurately and fast

Triaged support

When media talks about RPA and AI, it's implementation which is most talked about, and so support is a very important consideration many BAU teams new to RPA skim over or treat as an afterthought when looking to automate processes.

Understandably, for new RPA teams, it's assumed that if the developer built it, they should be able to support it. My research suggests that one developer/support engineer can support up to 7 to 12 automated processes of medium to low complexity. However, what if the bot runs 24 hours a day? What if several bots fall over at once—how do you prioritize which bots get fixed first? And if you're supporting client bots (internal or external), what will be the agreed SLA the customer needs to put into place?

From a client viewpoint, we discussed earlier the hidden costs of support. However, from a CoE perspective, there are many considerations to take into account before you are ready to fully roll out automation for various clients or various teams. Some clients may require (and thus be willing to pay for) 4-hour fixes, whereas some, especially in finance, may need full-time supervision during working hours. Prioritization which is paired with an SLA should be agreed upon at the start when a new bot is in the design stage. These support costs are what the business analyst will put on the business case.

Once the bot is fully supported, a ticketing system can manage incidents and requests to modify the bot, just like a typical IT department. A portal allows process owners and clients to raise a ticket, request a bot enhancement, or alert the support team of a defect in the outputs or a change to the service.

This, in ITIL speak, is *passive problem management*; however, your Support team can also engage in *pro-active problem management* where they are actively looking for issues in the code, the inputs and the outputs, and can report back to the process owner the issues and their plan on how to solve them.

Here's an example of a basic triaged support process:

i. IDENTIFICATION

    a. The BAU team raise a ticket to inform the support team that outputs of the bot are incorrect or that they have noticed an error or received a high number of alerts from the bot.

    b. As part of the support service, the Support team regularly check on the bot 1-2 times a day; this could be a few minutes at a time for straightforward processes. During this, the Support engineers may identify an issue (proactive problem management) and inform the BAU team of the issue.

Daily support may include checking input data for completeness or data integrity, checking that the bot and server are up and running, and ensuring the output data is correct and sending to the right place.

ii. INVESTIGATION

   a. Level 1 support does some initial investigation and may be able to fix the issue. The solution could simply be to rerun the bot. If L1 support cannot fix or find the error, the issue is logged and escalated to a Support Engineer.

iii. FIX

   a. If this is a defect in the bot, this fix may be included in the support agreement the BAU team has with the CoE.

   b. If this is a fault due to wrong data from the business, the engineer may suggest an enhancement that can be made to the bot so it will accept this new type of data, or suggest improvements on the front end to ensure only correct data gets through (this may require re-engaging the BA to provide some continuous improvements).

   c. If this was a request for a Change to the functionality of the bot, this can be forwarded to the developer to add these enhancements (at this point the BA would have already carried out her analysis and created a new business case to justify this new change).

iv. RELEASE

   a. No matter the type of change, the new code would need to pass through the Transition Manager's process and released in an agreed change window.

Bot portals and dashboards

In a more mature CoE, to save time, the portal may also allow process owners to make service changes themselves if the change doesn't impact the bot code or process structure but instead is just to change a variable. For example, if the BAU team need to change the file location from which the bot accesses data, or add a new entry to the list of variables the bot has (e.g. a list of email addresses), then the process owner should be able to do this themselves without going through the triage system.

Forecasting and re-scheduling

As mentioned, the support team holds the metrics on the bots and can see how all processes are scheduled, so the Service Delivery Manager needs to include volume, AHT, and other performance metrics in her weekly catch up with process owners and discuss any seasonal spikes or increase in volume the BAU team foresee in the near future.

As an RPA analyst in London, I worked closely with the Development team to create a dashboard to generate reports on volumes, AHTs, and exception rates of processes on a monthly basis. This saved the Development teams time, as this was a self-service dashboard. It allowed me to work with the BAU team manager to validate actual savings and accurately forecast exceptions, which enabled him to plan for the amount of man-hours his team would still need to dedicate to handling fallout from the bot as the company expanded and overall volume of work increased.

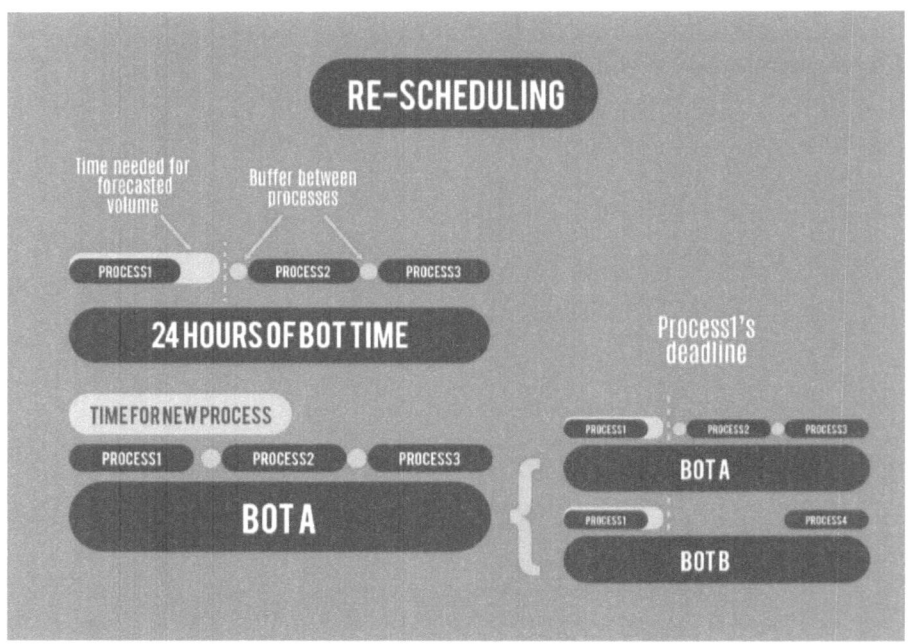

This insight meant we could work on solutions to potentially improve the bot and save his team more time. By also looking at predicted ramp ups of future volume of work, I was able to work with the Support team to ensure enough time was scheduled on the Bot so that the increased time to complete process 1 didn't delay process 2 (as depicted in the image above). If there was not enough space on the current Bot, we would need to use a second bot to manage the extra work and reduce the time to complete all cases before the deadline.

34 Managing changes

When making changes to the production environment, whether that's activating new automation bots or making changes to existing virtual workers, it is vital that this is done in a controlled manor. You need to consider what you would do if you needed to reverse changes because development took longer than expected, or it wasn't the best time is to deploy automation (e.g. if for retail it may not be a good idea doing this on Black Friday).

Releasing new bots

The management of releasing a new bot into production should start once the bot has been built by the developer. Just as the RPA business analyst liaised with the developer to hand over the process design document (PDD), the developer now liaises with the support engineer to hand over the SDD (solution design document). As your development team and Support team all sit together in the Centre of Excellence, coding standards and acceptance criteria should be agreed upon by, and consistent in, both sides. Even so, roles, responsibilities, and a gated process need to be defined to ensure a successful rollout.

Are you in control?

Let's go through these questions to confirm whether your Support team have all their bots under control and the team has a scalable structure.

- Do you have someone carrying out each role shown in the Support CoE structure (even if some of your team are wearing several hats)?

- Do you have a transition manager who is a separate person to the developer and support engineer?

- Do both the Development and Support teams use the same agreed coding standards, best practices, criteria, and knowledge base?

- Do you regularly provide process owners with bot health and performance reports and discuss seasonal or future volume increases?

- Do you have a prioritization mechanism and triage process for handling requests and defects?

- Are you continuously looking to optimise your CoE through the same lens as when inspecting a BAU team? (automate repetitive tasks, using self-service portal for clients and process owners).

Now that your Support team is humming and working in tandem with the design side of your CoE team, and you've embedded continuous improvement on your CoE and digital workforce, there's only one thing left to do…

35 Circle back to the beginning: AWARENESS & EDUCATION.

In the next chapter, we will see that this is what the 'U' is all about.

U: Uncover, Upgrade & Upskill

The 36th Step:

It's important that the CoE team keeps up to date with the market and an eye on the evolution of automation technology and new entrants. RPA is just the first step. Many companies are combining this with AI, using computer vision to scan and interpret files (intelligent OCR, aka ICR), ML, data lakes (DL), and Natural Language Processing/Generation (NLP/NLG) to ingest big data to find patterns, answer customer (or staff) queries, and solve problems.

BPM software has become a great way of tying all these together, and with so many options in an ever-evolving domain, you and your core team need to continuously upskill so that knowledge is kept in-house, and external consultants (like Lean IA) transfer knowledge to your business.

As you may notice, this methodology has done a full circle back to Aware (A) and Educate (E) but now for new technology, which can enhance your automation capability.

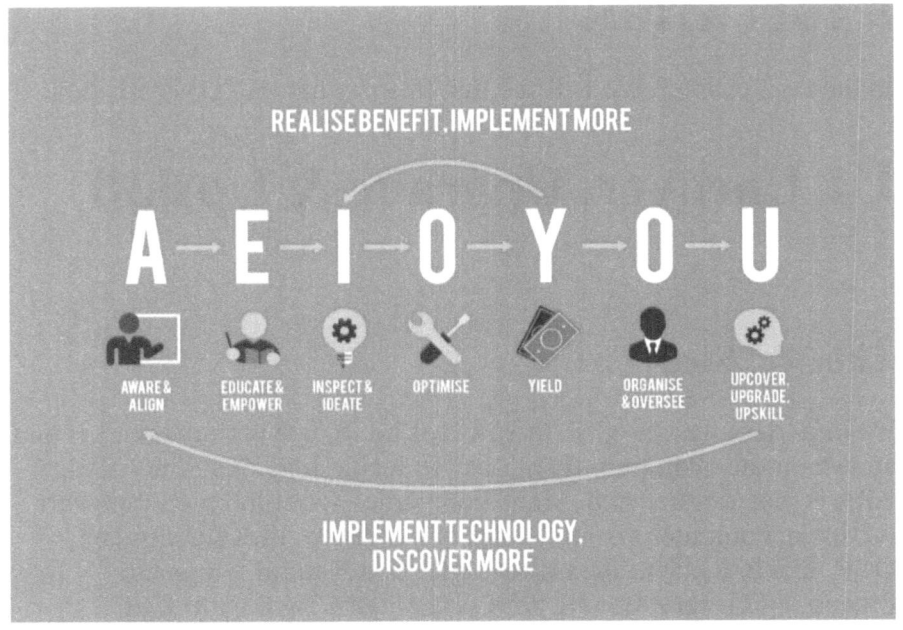

GUEST: Head of Automation and AI

AI is such an exciting topic and a trailblazing industry to be in. Full of potential optimism but also a lot of confusion as knowledge of how it works are held by a few. I met up with, **Gurpreet Ahluwalia, who's head of Automation and Artificial Intelligence at his company** *and has been working in AI consultancy for quite some time. I wanted him to share with you what insights he had on how to find the right solution and where this is all going.*

Great seeing you, and thanks for taking the time today to give us your insights. Could you talk a bit about your experience in AI?

My experience is more on strategizing induction of artificial intelligence for my company. I have been working across AI taxonomy primarily on machine learning, deep learning, NLP, vision and speech. Greater work is done in developing home grown and partner products on chatbots, virtual assistants, predictive and prescriptive analytics, object detection and recognition etc.

What attracted you personally to the AI/intelligent automation industry?

Innovative and disruptive technologies have always fascinated me to the core. AI was a hype some time ago before turning into reality. The challenges in the transformation journey, starting from the initial phases of the hype curve, have been very exciting. The scale and ways in which AI can propel business and IT leading to business outcomes is endless, which means enormous innovation, ideation, and incubation.

What struggles/frustrations and reservations have your clients expressed whilst trying to implement AI?

Top 3:

1. Talent – AI and data science has been from the 1960s. Given its sudden burst over the last 5 years, there is enormous scarcity of talent and human resource. While CSPs like Amazon, Microsoft, AWS, and Google have been trying to democratize AI for the citizen data scientists, it is still far from the point of operationalization.
2. Cultural and technology maturity:
    a. AI adoption is still nascent.
    b. Mid- and small-segment enterprises are still behind on the adoption.
    c. Specific industries like insurance are yet to reach acceptance. This is also applicable to segment use cases where accuracy is critical to business.

3. *Availability and sharing of data: data science and specifically machine learning / deep learning is all about data and its quality.*
   a. *Legacy systems have not been designed to store data for the purpose of harnessing.*
   b. *Specific industries like banking and financial services and insurance have had reluctance of sharing data for experimentation, to third-party providers, to software suppliers etc., which limits possibilities.*

What would be your top tips for matching AI solution/capability to the client's problem?

- *Start small with a POC followed by scaled implementations.*
- *Work in close alignment with business.*
- *Agree on acceptance criteria upfront.*
- *Technology adoption should align with the enterprise strategy for AI adoption.*
- *Work with real data – as close to production as possible.*
- *Focus on business value and outcome.*

Can you give a few tips for business leaders on how to find an AI vendor/technology partner?

- *Decide build vs buy. Partner channel devised accordingly.*
- *For majority of industry players, a combination of domain expertise and AI is the key – bilinguals.*
- *Look for operational experience in AI.*

- *Flexibility to support customization at right cost.*

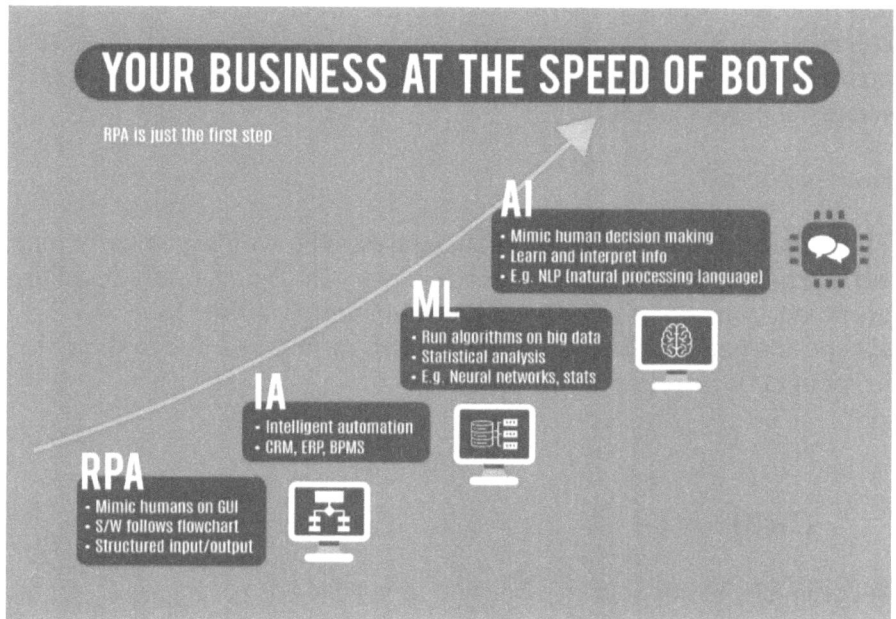

Let's give a brief non-technical/simplified introduction to an array of AI and intelligent automation solutions, '**Bots**' if you will:

Intelligent Automation, Analytics, and more

Process Discovery/Mining

What is it?

Process discovery is a subset of process mining, where software observes user actions on their computers to map steps a user makes in a process, mapping the various exception paths and even can identify trends and other patterns. This software can record process metrics such as volume, AHT, as well as how many applications and screens a process sues

How is it used?

This can save time and effort of an analyst who would manually map out a process, in addition process discovery tools accurately include every step, click and keystroke, and can identify all the long-tail exceptions paths an analyst may not have discovered due to their infrequent occurrence

CRM

What is it?

CRM (customer relationship management) is the management of

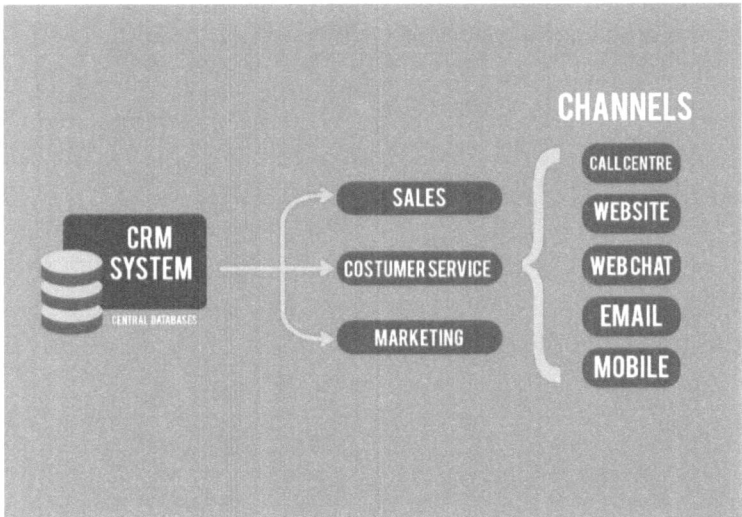

engagements a business has with its customer, and the management of customer data.

How is it used?

These CRM systems can collect a massive amount of data on a customer from their online activities on social media to their demographics and purchase history. Understandably, with so much big data, traditional analysis cannot fully extract all the value from this, so a strong trend in many CRM systems is to incorporate AI to discover customer buying behaviors and predict interests and make recommendations for your sales team to make informed decisions.

ERP

What is it?

Oracle defines it this way: "ERP stands for enterprise resource planning. It refers to a suite of software that organizations use to manage day-to-day business activities, such as accounting, procurement, project management, risk management and

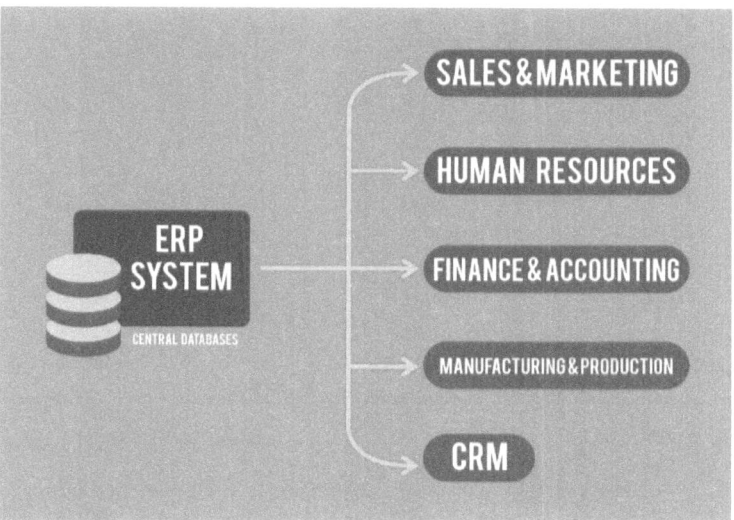

compliance, and supply chain operations".[9]

---

[9] https://www.oracle.com/uk/applications/erp/what-is-erp.html

How is it used?

Having a shared database gives businesses a single source of truth and allows information to flow between teams to provide better insights and enables the business to have seamless cross-departmental processes. Optimising these end-to-end processes when using ERP systems can reduce a business's operational costs, remove waste, and increase productivity. Similar to CRM systems, ERP collects a lot of data so ERP systems with embedded AI and automation can provide deeper insights into your internal business processes and your workforce.

BPM

What is it?

Business process management is similar to ERP, as this looks internally; however, BPM is more lower level and focuses on tying

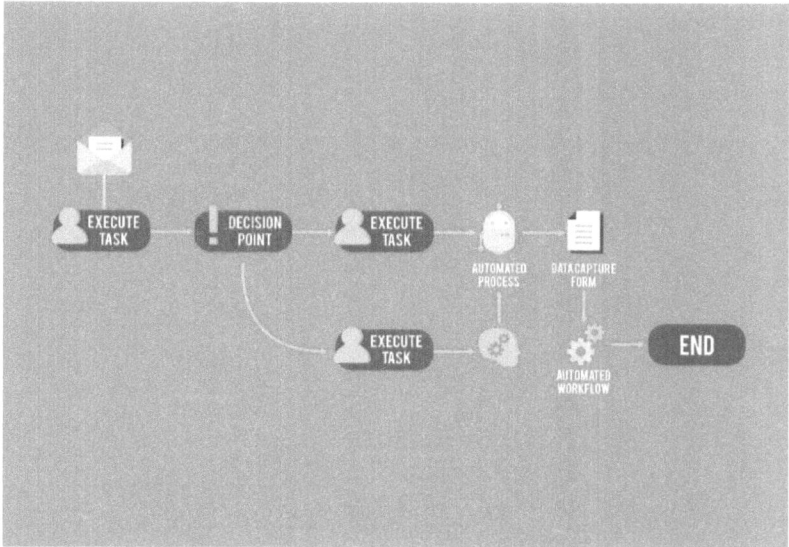

individual processes together, whereas ERP ties applications and departments together.

How is it used?

BPM can tie individual processes together using automated workflows, data capture forms, and RPA and IA bots to create a digital nervous system throughout your department or organization. The data it collects can provide powerful business intelligence (BI), and with embedded AI, the insights can be even richer.

Analytics

What is it?

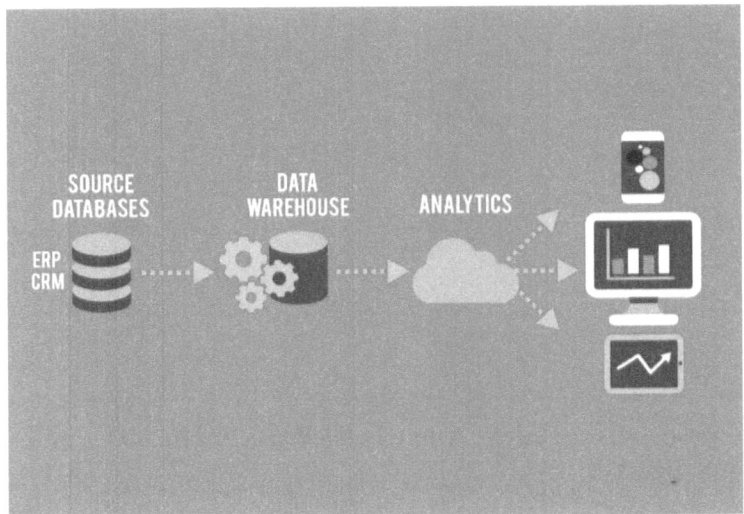

"Analytics is the scientific process of discovering and communicating the meaningful patterns which can be found in data. It is concerned with turning raw data into insight for making better decisions".[10]

How is it used?

---

[10] https://www.techopedia.com/definition/30296/analytics

Analytics is the end-point for your data collection systems such as ERP, CRM, or BPM. It visually represents your data in meaningful ways such as graphs, charts, and infographics to assist businesses in process optimisation, improving customer interaction, and identifying and minimizing wastes to help decision-makers make more informed decisions.

APIs

What is it?

"API is the acronym for application programming interface, which is a software intermediary that allows two applications to talk to each other".[11]

How is it used?

Though it's not AI or automation, APIs are used to allow applications to interact with each other and share data. Web APIs can extract information from web services. APIs can help streamline a process, though they can be difficult to build.

Artificial Intelligence

"A robot must obey the orders of humans except when such orders would conflict with the First law" - Asimov's 2nd law of robotics

Machine Learning

https://www.ibm.com/analytics/machine-learning

[11] https://www.mulesoft.com/resources/api/what-is-an-api

What is it?

"Machine learning is an application of artificial intelligence (AI) that provides systems the ability to automatically learn and improve from

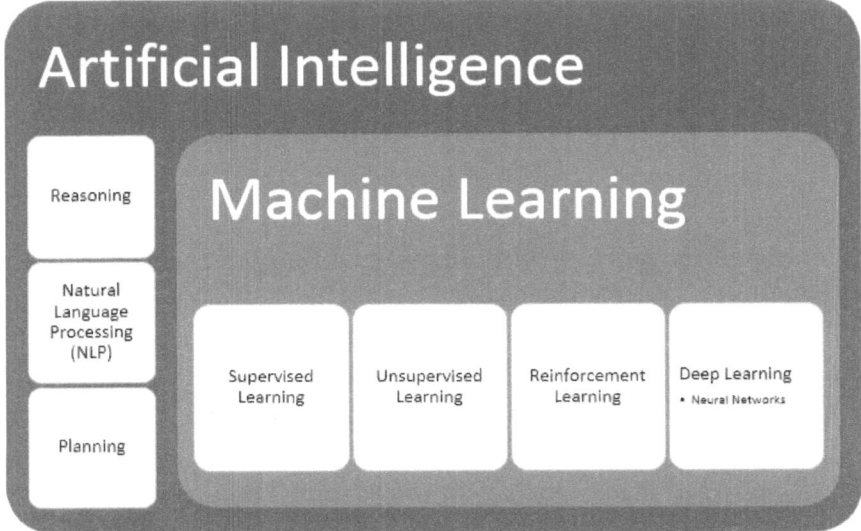

experience without being explicitly programmed".[12]

Machine learning is a subset of artificial intelligence where the machine learns from experience how to carry out a task to improve its performance or probability to get the answer right.

How is it used?

Supervised learning is where the algorithm is taught, by giving it a data set and the answers; then it's asked a new question to calculate what that answer should be. For example, an algorithm is fed information from a set of patients and told who does and doesn't have a specific illness. Then, the algorithm is given a new patient and calculates the probability that the new patient has the illness.

---

[12] https://expertsystem.com/machine-learning-definition/

On the other hand, *unsupervised learning* self-learns by ingesting a big set of data and discovering patterns in data.

Data Mining

What is it?

Also known as knowledge discovery in databases (KDD), data mining is "the nontrivial process of identifying valid, novel, potentially useful, and ultimately understandable patterns in data".[13]

How is it used?

It is useful in healthcare to detect trends and potential signs of illness early on. In marketing firms, data mining can segment customers into smaller groups to better recommend products and services. Financial companies can detect differences in purchasing behaviors to help detect fraud.

Computer Vision

What is it?

This allows machines to gain understanding from images and videos.

How is it used?

---

[13] https://medium.com/@exastax/the-history-of-data-mining-d2aeb0f587ce

It is used for image classification, facial recognition, as well as object detection in videos. Insurance companies can use it to automatically process images sent in by a claimant to detect the vehicles involved and the type of damage. Businesses can use it to process scanned images like a handwritten document or facial recognition to identify customers via their smartphone.

Deep Learning

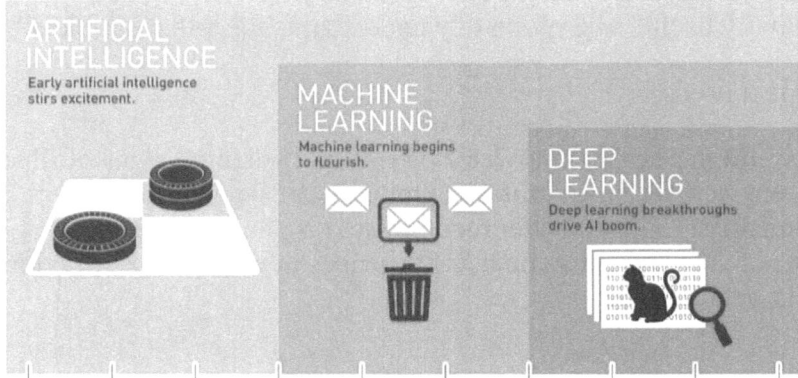

https://blogs.nvidia.com/blog/2016/07/29/whats-difference-artificial-intelligence-machine-learning-deep-learning-ai/

What is it?

"Deep learning is a subset of machine learning where artificial neural networks, algorithms inspired by the human brain, learn from large amounts of data"[14]

How is it used?

---

[14] https://www.forbes.com/sites/bernardmarr/2018/10/01/what-is-deep-learning-ai-a-simple-guide-with-8-practical-examples/#3cdf06b28d4b

Aside from using deep learning for computer vision, here are some interesting applications for deep learning:

Navigation for self-driving cars, recoloring black and white images, predicting the outcome of legal proceedings, precision medicine, automated analysis and reporting, game playing.[15]

Data Lakes

What is it?

Unlike a data warehouse, which is a massive storage of structured data, a data lake is unstructured raw data in its original form used by data scientists for AI engines to ingest in order to find patterns and make predictions.

How is it used?

This is just a big data repository storing data in its original form until the data has a defined purpose

Natural Language Processing/Generation

NLP is an area in which I will personally be spending the next few years focusing, as I believe it is central to the AI industry, as it encapsulates all forms for AI. This is because NLP is how we communicate with machines.

---

[15] https://devisionx.com/technologies/deep-learning/

Computer vision uses NLP to categorize items in an image to describe what is happening in a picture or video.

NLP uses machine learning to analyse texts in 'vector space' to make comparisons to determine the answers that are relevant to a question and then improve on the feedback they receive.

GUEST: *Artificial Solutions*

Sales Director at Artificial Solutions, Frank Burnett-Alleyne *was kind enough to provide a little insight into his world of highly intelligent Chat bots, with a closer look at their product,* **Teneo.**

So Frank, I've seen Teneo first hand and personally love the capability, tell us about what makes Artificial Solutions stand out in the NLP space, and Teneo different from other 'Chatbot' products?

Artificial Solutions was founded in 2003, making us a true trailblazer in conversational AI space. Our vision was (and remains to this day) that it should be possible for an ordinary person to have a conversation, in their native language, with a device or computer program and for that device or program to understand the request(s) and successfully execute it/them. We were at the forefront of developing natural language applications. We began as a consultancy staffed with computational linguists, neuroscientists, data scientists, and programmers because, unsurprisingly, there were no tools available to develop the natural language applications fundamental to realising this vision.

Teneo empowers organisations by providing an environment that does not require highly specialised skills in computational linguistics or neuroscience to create sophisticated conversational applications. External- or internal-facing, a Teneo application, whether a chatbot, virtual assistant, or concierge, can be deployed across the entire enterprise and surfaced through any digital channel—voice, text, web, or smarthome device—without codebase changes.

Our USP is that Teneo is the only GDPR-compliant platform with which it is possible to develop conversational AI applications, with or without training data, that can be surfaced across any digital interaction channel without code-base changes. Ownership of the conversational data held in Teneo rests with the organisation, not Artificial Solutions—our clients have full and unfettered access to their Teneo data.

Teneo overcomes the challenge of understanding the end-user's intent, however expressed, by seamlessly combining syntactic and machine learning techniques. Teneo's comprehensive intent recognition capabilities handle a range of areas including context, synonyms, relevance, and tense, understanding, for example, the difference between "You cancelled my policy" and "I cancelled my policy".

Teneo's integrated data tools capture all conversational data and associated metadata in real time, helping drive dynamic, personalised conversations with the end-user. Teneo provides a complete audit trail of the entire conversational journey and tools with which the business can quickly and easily maintain and optimise the conversational flows. The verbatim voice of the customer data captured by Teneo is surfaced as unaggregated data, enabling the business to derive valuable insights, identifying emerging trends and allowing evidence-based decision-making.

Conversational applications developed with Teneo go beyond answering simple FAQs, expanding to include automating complex, multi-step processes and transactions, all of which can be accessed through the end-user's digital channel of choice. With Teneo's capabilities, including the normalisation of conversational data to enable consumption by RPA systems and support for synchronous and asynchronous RPA processes, Teneo helps maximise opportunities for automation by extending the reach of RPA programmes.

AI is this new and exciting, cutting-edge market, and I believe NLP will be at its center, but what attracted you personally to this industry?

I was attracted to NLP/AI because, after 10 years working in the contact centre industry, it was increasingly evident that a significant proportion of end-customers would not shift channels, and that a rapidly growing proportion expected personalised, frictionless, and immersive experiences, delivered consistently using their interaction channel of choice. NLP and AI appeared to be the technologies with the greatest potential to successfully address the needs of both organisation and end-customer, and with my background it was a natural fit.

Have you noticed any trends in the chatbot arena and ways it's impacting your clients? Where do you think this is going?

The market is evolving at breath-taking pace, with many organizations now offering chatbots capable of answering simple FAQs.

An emerging trend is that end-customers are expecting chatbots to do far more than just answer FAQs, wanting to interact with more sophisticated chatbots, capable of successfully handling complex queries, in the end-customer's interaction channel of choice. They don't want to channel-shift.

Teneo empowers our clients, allowing them to embrace this market evolution by creating conversational AI centres of excellence to satisfy the emerging demands and establish and extend competitive advantage.

What we are seeing, particularly in the financial services domain, is an increasing desire for voice-driven chatbots. I don't mean chatbots that make outbound calls; that's relatively easy to implement, as the chatbot is effectively controlling the scope and domain of the conversation. The vision, which is more challenging, interesting, and mutually beneficial to end-user and organisation, is using chatbots to seamlessly handle inbound calls. With inbounds, the chatbot is not controlling the conversational scope and so is reliant on its natural language interaction capabilities (our term for combining natural language understanding, processing & generation) to understand the end-customer's intent(s), however expressed, and successfully guide them through a conversation to reach the desired outcome.

This outcome could be the successful resolution of the end-user's query or dynamic guidance through a process that previously involved tedious, procedural form-filling. Conversational AI can reduce friction in the end-user experience. By assisting customers at certain points in a transactional conversation where a web form cannot, it helps reduce drop-out or basket abandonment, thereby improving conversion rates of sales/transactional processes. The benefits of intelligent conversational AI are immense: the service is available to the end-user 24/7, the organisation is engaged with the end-customer—critical in these times of diminishing customer loyalty—and secures valuable insights into end-customer behaviour and the organisation's operational effectiveness. Personalised, intelligent automation helps makes the experience more relevant and processes more efficient and enables organisational scaling.

We are about to go live with a customer deployment in the FS sector that is combining voice and web channels to drive a multi-step process that mirrors what I've described and represents the art of the actual, not the art of the possible.

How is/will NLP/conversational AI take centre stage in the AI and intelligent automation world?

NLP/conversational AI is taking centre stage in the intelligent automation world through its potential to extend the reach of RPA programmes. The successful execution of sophisticated, multi-step processes that Teneo conversational chatbots successfully automate are, by their nature, rich in data. Teneo's in-built functionality makes it quick and extremely easy to normalise this conversational data so that it can be consumed by RPA systems, and for Teneo to consume the resultant output from the RPA system to drive the conversation.

What are the biggest/most common struggles and frustrations companies are experiencing when trying to find and implement the right chatbot/NLP solution in a crowded market that meets their unique needs?

Sadly, there's a lot of hype in this domain. Machine learning is an exceedingly powerful capability, but the fact is that for it to work, the organisation needs training data, and typically lots of it. Machine learning is still mostly a "black box" solution, which means little or no control over tuning the conversational AI to understand subtleties or nuances. One frustration is that the first release of a chatbot typically understands very little, sometimes as low as 20%. This is because the chatbot is learning, and the learning takes time and lots of data—but clean, curated data that typically requires significant effort to prepare. The consequences are frustration on the part of the end-user and the organisation, as 20% levels of comprehension means many escalations to agents, generating a level of failure demand that wasn't in the business case, lower CSat ratings, and the likelihood of increased customer churn.

The Teneo platform deliberately combines machine learning with syntactic techniques to add refinement and breadth. This allows Teneo to understand conversational nuances and subtleties, which results in intent recognition and comprehension rates that in the first release are typically above 70%, and very high levels of containment. Teneo's iterative process of conversational flow optimisation continually improves on those figures.

Another frustration is when organisations discover that, the tools selected because they allow very rapid chatbot creation under evaluation conditions, are sadly lacking when it comes to building, deploying, optimizing, and maintaining chatbots for production release. Digital interaction channels are not just convenient, they are also a rich source of data that can surface insights, intelligence that should be made available to the organisation to help them improve across all areas. The reality is, few of the tools used to build conversational applications facilitate, out of the box, capture of this voice of the customer data, easy interrogation of the data or access to all of it, unaggregated, so they can generate insights. Many organisations have built and deployed chatbots yet have no way of measuring how they are performing, or understanding what they are doing well and where, what they could improve and why, what their end-customers think of them, and what they would like them to offer in future.

The organisations that think about what benefits they want to realise from conversational AI; when, why, and what they want to achieve once they have launched a chatbot; and how they want to get there will find it easier to navigate the options available.

Common to all our clients is their desire to have control and agility, and to ensure they have the headroom to grow.

With all these exciting new technologies readily available, it can be frustrating for you if you're a small business; you're probably feeling a little left out right now...

Cloud Computing

What is it?

Microsoft Azure says, "Simply put, cloud computing is the delivery of computing services—including servers, storage, databases, networking, software, analytics, and intelligence—over the Internet ("the cloud") to offer faster innovation, flexible resources, and economies of scale".[16]

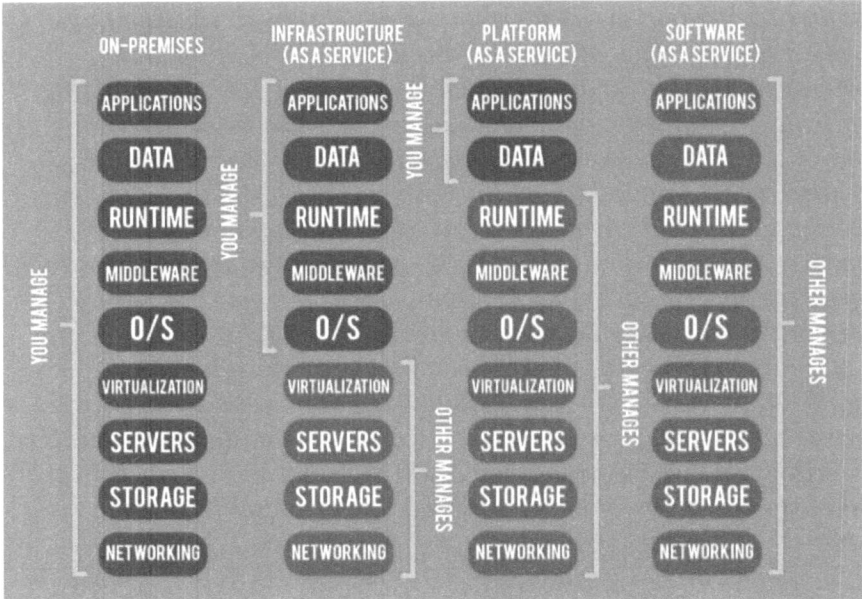

How is it used?

Cloud computing allow businesses to use applications, services, and even processing power or storage without high upfront costs, as there is no need to purchase hardware or software outright; you can pay for what you use. Cloud services also are useful for collaborative working, where several colleagues can simultaneously work on the same document.

https://www.bigcommerce.co.uk/blog/saas-vs-paas-vs-iaas/#the-key-differences-between-on-premise-saas-paas-iaas

These different types of cloud-based services come in the form of SaaS (Software as a Service), PaaS (Platform as a Service), and IaaS (Infrastructure as a Service), best depicted in the picture above.

---

[16] https://azure.microsoft.com/en-gb/overview/what-is-cloud-computing/

Super-charge your bots to do more

Incorporating AI capabilities to enhance your RPA bots can open up a world of opportunities beyond just automating the manual, repetitive tasks.

Real world example: ICR and data lakes pilot

Situation

In the world of financial advice, many IFAs (independent financial advisors) and their clients still prefer paper and wet signatures as opposed to electronic documents. This causes issues for many wealth management firms, as their staff may need to scan the documents for audit purposes and also need to manually type the details into the system. This is a long-winded process, can take a lot of time to process one document, and can result in errors when staff have to repeatedly transfer information from a pile of paper into the system.

Outcomes

Using an AI engine which included intelligent character recognition (ICR), a data lake, and machine learning, we were able to have the engine ingest the scanned images, set certain conditions for different fields in the document form (e.g. country, city), and feed the resulting data directly into the database.

Using high-quality printed documents and scans, the engine was able to understand both text and handwritten information, and with machine learning and the set conditions, it was able to scan the words and intelligently interpret the content (e.g. LDONON in the city field was automatically changed to LONDON). This reduced the average handling time and data input error (as the engine would only input information which was above a high probability threshold), and provided scalability.

Real world example: RPA and chatbots concept

Situation

In the financial industry, especially in asset management, clients want fast responses to their queries. Speaking with operators in the client services teams, my team discovered that the answers to many queries already existed somewhere on the company's website, or in the client's profile. Furthermore, staff expressed that they found they were carrying out similar transactions time and time again (e.g. requests about their portfolio). This can take new staff time to find the right information, and repetitive activities are always prone to inconsistency and errors.

Outcomes

The concept proposed and prototyped was to create a knowledgebase of frequently asked questions and answers compiled by the most knowledgeable and experienced of the company's workforce, a database of information from the company website, and by building automated processes with RPA of the most frequently requested transactions on client accounts. This could then be combined with a conversational chatbot so users would be able to ask the chatbot a question which would either provide an answer or activate an automated transaction.

This also demonstrated how new staff could get the right answers by interacting with the internal chatbot so they could learn and provide a consistent service. Customers could request information or make asset management requests 24/7 and get accurate responses fast.

Have you identified ways to enhance your bots to gain more benefits?

Now that your digital workforce and bots are in full force it's time to make sure your CoE team keeps circling back to the beginning of the AEIO YOU lifecycle to stay aware of new emerging technology and education.

- Is your team aware of the plethora of different types of AI capabilities on the market?

- Have you meet with different AI vendors and watched demos of what their products are capable of doing for your business?

- Have you run any POCs with vendors to demonstrate first-hand how much value their 'bots' can add to your CoE team?

- How serious (or nervous) is your business at implementing AI into their business processes?

Let's take a recap of the YOU chapters:

- Yield: Have you realized the benefits to show the value of your RPA bots?

- Organise: Are bots smoothly being transferred from the development to the support team, and is your support team maintaining the continued running of your bots with regular checks and an efficient triage process?

- Oversee: Is your support team managing bug fixes and modifications in a controlled manner?

- Uncover: Are you constantly discovering and learning about new technology which you can incorporate in your automated processes? Is your team aware and aligned with how you can use this new technology?

- Upgrade: Are you identifying and assessing potential use cases and opportunities for this new technology?

- Upskill: Is your team gaining new skills and expertise in new technology that may be introduced into your organization in the near future? Are they educated and empowered, ready to take on this new technology?

GUEST: Operations Director

Steve Waldron *has been* **Head of Operations** *at several companies in the financial sector in London for over 30 years. I enjoyed discussing with him about his experiences and his opinions of the future of finance*

Hi Steve, so what attracted you personally to fintech and the use of intelligent automation?

I have been in the financial sector for over 30 years, working for different organisations across the sector in various operational and client service/relationship management roles. I have always been a fan of 'disruption', so when the world of 'Fintech' came along I was excited. Having been a Head of Operations a few times now, there can be real frustration with the level of manual processing in an Ops department. Fintech meant an opportunity for an Ops department to become more automated, scalable, reduce risk, reduce cost and invariably help generate more revenue.

The great thing about Fintech is that it covers a number of exciting initiatives such as RPA, AI, Machine Learning and Blockchain. For an Ops department and an Ops Manager with vision, this is golden.

From what you've experienced, how do you see Ai/RPA impacting the financial market, and what value does it bring?

AI and RPA have been in the financial sector for quite some time now, as with other sectors too. The reality is some people still do not realise that Artificial Intelligence and Robotics are already part of their daily lives in their workplace.

I see a very positive impact for the financial sector because from experience the big investment banks usually lead the change to implement new ideas and technology, partly driven by their customers demand and partly because they want to be pioneers and early adopters - this will help them reduce their cost and increase revenues. The value this can bring to any organisation is endless, with the right vision from the decision makers this can be revolutionary in any organisation.

What limitations have you experienced with 'RPA' and struggles organisations have with adopting/implementing Ai?

Any limitations I have experienced with the adoption and implementation of RPA has been more to do with people and internal politics, by that I mean the development and cost of an RPA has been too high, which is always a challenge when getting the project signed off. The other challenge which is invariably an ongoing challenge for any Head of Operations is to convince people that this does not mean redundancies, so sometimes this is about 'hearts and minds' PR.

Another big challenge is multiple systems and processes which do not talk to each other, trying to integrate any of these using RPA and then feed into a core platform has been incredibly challenging, usually because different internal stakeholders have different opinions of RPA and automation in general and can become quite territorial about their world.

How have organisations you've worked with discovered and vetted new intelligent/AI tools/software?

If you work for an organisation where you are in a satellite office then this can be a challenge when it comes to discovering any kind of new AI type software, because the head office has vetted and selected what software to use which may not be the right choice for you in a satellite office.

Sometimes your client base introduces you to vendors who are in the AI space because your clients use them or they see an opportunity where they and you will benefit by using AI. If you are open minded and comfortable with change and disruption you will survive - this is still an issue today.

Where do you think RPA/intelligent automation market is going and how will this impact the finance and fintech sectors?

This is a good question - data is the key driver to all of the change, all organisations in the financial sector are sitting on a mountain of data, client data. How you use that data to your advantage is key.

Virtual Analytics will play a big part in the future too. Artificial Intelligence, Machine Learning, Robotics, Blockchain, Intelligent Automation are all being used somewhere in the financial sector.

Data is the new currency......If you look at any Ultra High Net Worth clients of any Wealth Managers or Asset Managers, there will be a big shift of generational wealth soon, probably already happening today. These inheritors of this wealth live in a different world to their parents. They are very tech savvy, have had an online profile for years, will be on many social media platforms and will have a social mindset. Everything for them is about convenience, one touch button mindset and will expect to see everything in one go on their smartphones or ipads.

They will expect to see their portfolio live at any time of the day in any location around the world on their smartphones, they will want to have conversations with their wealth relationship manager on Whatsapp or WeChat they will not want to go to a face to face meeting in Mayfair.

So, the visionaries at any financial organisations, the CTOs or CTIs will be instrumental in ensuring their organisations are ahead of the curve by pushing their boardrooms to invest and implement the different types of opportunities within the world of 'Fintech' because they have the data - how they use their data will determine if they survive the new era of disruption or not.

The 36 steps of the AEIO YOU method

We're grateful that you've read this book and taken it with you along your intelligent automation journey. Hopefully you've learnt a lot, found the insights of the contributors were valuable, and this book has answered your most pressing questions.

If you've followed these steps, we hope you've saved a lot of headache and brought your CoE team to a maturity level to proudly call it an Automation Factory. Now start the book again from step 1, but this time with a newer technology.

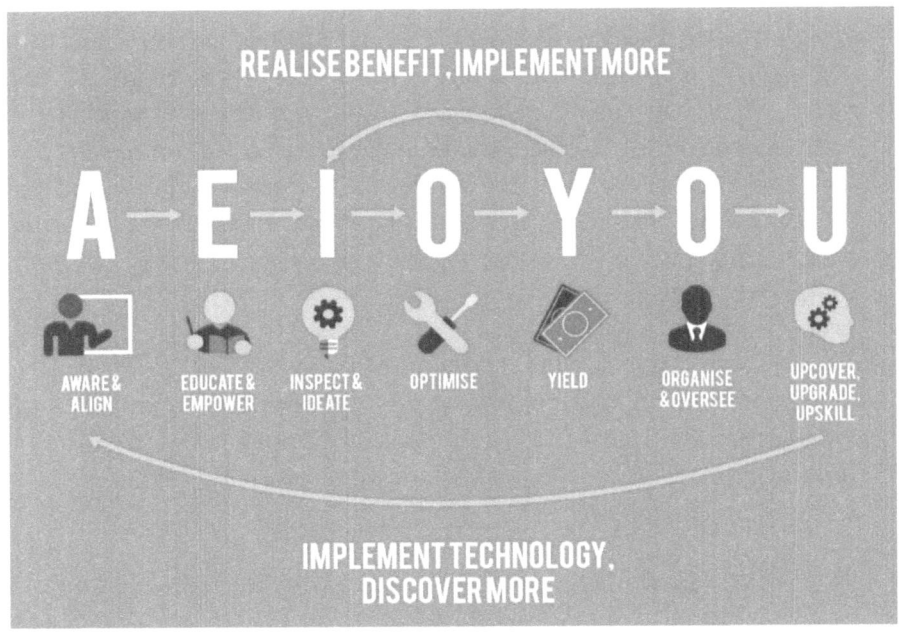

Here are the 36 steps of the AEIO YOU method:

        A

1. Understand the Technology
2. Know the myths, Challenges and the Benefits
   2b. Know the common mistakes and pitfalls
3. Understand the market
4. Choose the right solution provider
5. Be an evangelist
6. Run a Pilot (POV rather than POC)
7. Start building your Centre of Excellence

        E

8. Bring in the Experts
9. Involve staff so they welcome the change

10. Keep hold of your most valuable assets
11. Upskill to build capabilities inhouse

I

12. Zoom out. Create an Enterprise roadmap
13. Zoom in: Define and Measure the problem
14. Filter on what's suitable
15. Focus on the top 20% (80:20 rule)
16. Build a complexity map: Cost vs Benefits vs Financial savings
17. Prioritise: Go for some quick wins first
18. Zoom in further: Business Analysis 101 (Data and requirements gathering)
19. Root cause analysis

O

20. Solution design
21. Lean thinking
22. Business Case: detailed cost-benefits analysis
23. Define the process
24. Design the solution
25. Build the solution using best practices
26. Prepare the data for testing
27. UAT
28. Hand over to support team
29. Launch
30. Reflect
31. Repeat and scale

Y

32. Realise the benefits

O

33. Maintain the benefits
34. Manage the changes

U

35. Circle back for continuous improvement
36. Discover newer technologies

The next steps:

Step 1: Take the RPA team diagnosis test https://leania.co/rpa-team-diagnosis/ **to get an 11-page personalized report of your team's readiness and CoE maturity scored against each section of AEIO YOU**

Step2: Request a tailored Centre of Excellence Operating model pack https://leania.co/the-coe-starter-kit/

Step 3: Access our Training portal to train your team on the AEIO YOU method www.leania.co/leania_academy

FINDING THE RIGHT AI or Intelligent automation solution

The 36 steps of the AEIO YOU method can be applied to implement any new technology in order to further optimize your business towards full digital transformation

Finding the right AI solution is a tedious feat of reading through white papers, watching videos, sitting through many pitches and demos whilst you keep asking the same questions over and over. Why not just send us your problem statements or use-cases and we can quickly recommend which products and providers can solve your unique needs.

The AI and automation finder

Feel free to tell us exactly what you are looking for, without limit

robotic process automation

FIND

Find the right AI and automation tools that will match you business/ use case

COMPARE

Find the right AI and automation tools that will match you business/ use case

DISCOVER

Discover new technologies that can enhance your digital transformation strategy

Acknowledgements

After speaking with directors, heads of department, RPA developers and analysts in and around central London, I kept hearing the same myths, challenges and pitfalls they or their clients were making. Many companies I had spoken with and companies online were repeatedly making similar mistakes, so I decided to write this book. I reached out to Guy Kirkwood, chief evangelist of UiPath to see whether he would allow me to share some of his insights and experiences in this book which he warmly agreed to. His support encouraged me to reach out to many other business leaders, experts and CEOs in RPA and AI.

Always keen to discuss automation and its future over a coffee, my passion for this sector and my career has led to meet many interesting people working on exciting new technology, some of whom are featured in this book.

I'd like to thank; UiPath chief evangelist Guy Kirkwood, Teneo sales director Frank Burnett-Alleyne, Mimica automation CEO Tuhin Chakraborty, SoftBot CEO Mike Beason, Co-founder of Edge Tech Harrison Goode, Solutions Architect David Orton, Tabscanner CEO Rashad Al-Safar, Broadgate consultancy partners John Vincent and Richard Gale, public speaker and director of Wzard Innovation Rob King, head of automation and AI Gurpreet Ahluwalia and Operations Director Steve Waldron

I would also like to thank my editor George Wilson and he team, as well as Stenio Santos for the artwork

Finally I would like to thank PEX editor Ian Hawkins who reached out to me for his article which allowed me to share my thoughts of intelligent automation with their 150,000+ members worldwide

About the Author

Tony Walker is Founder and CEO at Lean IA Limited, who, leaning on his experience and seasoned network of experts designed the AEIO YOU methodology, to be a easy-to-remember, easy-to-follow step-by-step process to help businesses digitally transform.

A Lean Six Sigma Black Belt, an RPA Business Analyst/Project Manager by trade, though an Aerospace Engineer and Financial Analyst by training.
He's helped set up or mature both Design and Support CoEs and worked with some of the world's largest consultancy firms, to help client's get bots up and running, and ensure that they run smoothly with the right controls.
Tony has worked in Insurance, Aerospace, Utilities, Fintech and Asset Management firms and is always keen to share his and his network's experiences, insights and visions of the future.

A bit more about the story behind AEIO YOU:
The ethos is simply 'Empower teams transform businesses'

Our mission is to empower employees to discover and use new technologies in order to become self-reliant at optimising their own organisation, by giving them the know-how as we work with them to automate with RPA and IA.

We want to share knowledge to assist businesses new to RPA and IA so that they don't make the common mistakes of those who have gone before them. We believe educating and developing teams is the way forward

We do this by eradicating poor success levels seen in the RPA and AI industries by tackling the 3 root causes we have discovered for this. This technology is AMAZING so why do 50% of RPA and AI projects fail, and less than 5% of businesses have successfully scaled.

Our values are: EMPOWER, OPTIMISE AND DISCOVER
We provide guidance, resources and web applications to you, the people who strategise, implement and manage the automation; the Analysts, Project Managers, Service Delivery and Change Managers
Through our own journey of RPA and AI, implementing governance and automation, and educating teams, we realised that though there are lots of resources, guidance and great courses for developers online, there still is hardly anything out there for educating teams who govern and deliver intelligent automation
We believe automation should be on a lean process, and so our approach is heavily based on lean thinking, and Lean Six Sigma
We want AEIOYOU.info to be the place for new CoE teams to find support on implementing intelligent automation, but also to support their continuously improvement

You may be feeling inspired to learn more, so what's the next steps? Keep in touch to stay up-to-date on new insights, events and tools to assist you and your team.

Here are 3 final steps to get your team going:

Step 1: Take the RPA team diagnosis test **to get an 11-page personalized report of your team's readiness and CoE maturity scored against each section of AEIO YOU**

Step 2: Access our Training portal to train your team on the AEIO YOU method

Step 3: Request a demo of the LIA web-app, the automation opportunity analyser

Acronyms

AEIO YOU: A method for successfully implementing intelligent automation into an organization
IA: Intelligent Automation
RPA: robotic process automation
FTE: Full time employee
AI: Artificial intelligence
GUI: Graphical User Interface
SME: subject matter expert
PDD: Process Definition Document
SDD: Solution Design Document
SLA: Service Level Agreements
API: Application Programming Interface
ICR: intelligent character recognition
COE: Centre of Excellence
CBA: Cost benefits Analysis
RCA: Root cause Analysis
A case: a unit of work that flows through a process, e.g. an individual customer request
PMO: Project/programme management office/officer
UAT: User Acceptance Testing
POC: Proof of Concept
POV: Proof of Value
EAR: Enterprise Automation Roadmap
MPV: Maximum Potential Value
TPV: Target Potential Value
AHT: Average handling time
SIPOC: Supply, Input, Process, Output, Customer
COO: Chief operations officer
TT: Takt time
GDPR: (General Data Protection Regulation)
CASS: (Protection of Client Assets and Money)
ML: machine learning
OCR: Optical character recognition
BPO: Business process outsourcing
BPM: Business process management (or Mapping)
VM: Virtual Machine

VDI: Virtual desktop interface

www.ingramcontent.com/pod-product-compliance
Lightning Source LLC
Chambersburg PA
CBHW021356210526
45463CB00001B/120